given to F
by Alec Cr

CW01082303

DAVIDSONS OF MUGIEMOSS

DAVIDSONS OF MUGIEMOSS

A History of C. Davidson & Sons Makers of Wrapping Papers, Paper Bags, Paperboard and Plasterboard Liner

J. NEVILLE BARTLETT

THE ATHLONE PRESS
London & Atlantic Highlands, NJ

First published 1997 by
THE ATHLONE PRESS
1 Park Drive, London NW11 7SG
and 165 First Avenue,
Atlantic Highlands, NJ 07716

© J. N. Bartlett 1997

British Library Cataloguing in Publiction Data
A catalogue record for this book is available from the British Library

ISBN 0 485 11514 X

Library of Congress Cataloging-in-Publication Data

Bartlett, J. Neville.
Davidsons of Mugiemoss: a history of C. Davidson and Sons makers of wrapping papers, paper bags and boards/J. Neville Bartlett.
p. cm.
Includes bibliographical references and index.
ISBN 0-485-11514-X
1. C. Davidson & Sons—History. 2. Paper products industry——Scotland—History. I. Title.
HD9831.9.C2B37 1997
338.7′676′0941—dc21 97-1025
CIP

All rights reserved. No part of this publication may be reproduced, stored in a retrieval system, or transmitted in any form or by any means, electronic, mechanical, photocopying or otherwise, without prior permission in writing from the publisher.

Typeset by
RefineCatch Limited, Bungay, Suffolk
Printed and bound in Great Britain by
Cambridge University Press

Contents

Preface

Some months after I had taken early retirement from the Economic History Department of Aberdeen University together with over a hundred colleagues as a result of government cuts, Colin Maclaren, the University Archivist, informed me that the management of C. Davidson & Sons were considering commissioning a history of their firm. As I had recently written articles on the history of two other papermaking firms in the Aberdeen area he thought I might be interested.

When I visited Mugiemoss to discuss the project I was at once impressed by the determination and enthusiasm of John Goodall, then mill manager and now Chairman and Managing Director of BPB Paperboard Ltd based at the Group's Northwich office. He launched the project, assured me that I would be given a completely free hand in my research and writing, and has given me invaluable advice and support.

The content of the history, of necessity, reflects the sparse and intermittent nature of the material in the firm's archives. I cannot claim that the surviving records, like those of William Morris the car maker, are 'vestigial in the extreme' but the very meagre data on such important matters as the composition of output and investment and on prices and markets for products, plus a woeful deficiency of personal and business letters, directors' minutes and memoranda which would shed light on policy decisions, has made my task somewhat difficult. Only one directors' minute book, covering barely two years, has survived from 1875 when the company was formed until 1948, whilst Colonel Davidson's Letter Book covering a few months of 1940 and 1941 is the sole collection of letters giving a detailed and regular picture of any aspect of the firm's operations. I persisted in my task because I felt that

C. Davidson & Sons had played an important part in the wrapping and packaging sector of the British paper industry, a sector which has been relatively neglected by historians in favour of more glamorous sectors such as high quality writing and printing papers or newsprint, where powerful press barons dominated the scene.

My debt to the staff of various libraries and archive collections is considerable. Their assistance was particularly appreciated in that it was often achieved despite financial stringency which entailed reduced manpower and resources. I should like to thank the staff of Aberdeen University Library, and the National Library of Scotland, Manchester Central Library, and the British Library Newspaper Library, Colindale for their patience and kindness. However I was particularly grateful for the assistance of Jennifer Beavan of Queen Mother Library and to Colin Maclaren, Ian Beavan and Myrtle Anderson-Smith of the Department of Special Collections and Archives at Aberdeen University.

At Mugiemoss itself detailed advice and assistance was given by Arthur Brown, Administration Manager, until he retired at the end of 1988. Since then Murdo MacDonald, formerly Commercial Manager and now mill manager, has performed a similar role, maintaining the channels of communication and ensuring that any minor problems arising after I moved to Northallerton and used this as my base for research and writing, were speedily resolved. His secretary, Lesley Flight, has been responsible for typing the manuscript undaunted by my handwriting or by late additions and amendments. Without their help my task would have been much harder.

On a more personal note I have been encouraged by the interest shown by former students of mine at Dudley and Aberdeen. Even more I am thankful for the patience and understanding shown by my wife.

Northallerton, 1995 J. Neville Bartlett

Tables

Illustrations

Units of Measurement and Value

(Decimal/Metric Equivalents)

1 pound (£)	=	20 shillings (s)	=	240 pennies (d)
1 shilling	=	12 pennies (d)		

1 foot	=	12 inches	=	30.48 centimetres
3 feet	=	1 yard	=	0.914 metres
1 mile	=	1760 yards	=	1.609 kilometres

1 ton	=	2240 pounds (lbs)	=	1.016 tonnes

Chapter 1

The Early Days

CHARLES DAVIDSON'S NEW VENTURE

In the second half of the eighteenth century as the tide of economic expansion quickened, Aberdeen emerged as one of the centres of the rapidly growing Scottish paper industry.[1] The city possessed several attractions for the papermaker. The city's lawyers, the staff and students of King's and Marischal Colleges, and the *Aberdeen Journal*, which began as a weekly newspaper in 1748, together with a growing population active in industry and commerce, generated a healthy demand for paper products. The basic raw materials for paper making, waste linen and cotton rags, as well as old ropes and sails, were readily obtainable in the city. Further supplies were shipped to Aberdeen from elsewhere in Britain and the Continent, whilst an additional source of supply was provided in the later years of the century by waste from the newly founded flax spinning and cotton spinning mills.[2]

An ample supply of clean water for the various processes of paper making was at hand in the rivers Don and Dee and their tributary streams. However an equal attraction was the water power they provided which could convert the waste rags into pulp, the only stage of paper manufacture which was mechanised successfully until the 19th century.[3]

No fewer than four paper mills were erected in the Aberdeen area during this period. The first began operations in 1751 beside the Culter Burn, which runs into the Dee seven miles west of Aberdeen and was erected by an English immigrant, Bartholomew Smith. In 1771 another mill was started at Stoneywood on the Don four miles from Aberdeen by Alexander Smith a local wigmaker who was not related to his name sake. Six years later he had a second mill on the same site. When Alexander Smith died in 1796 the Stoneywood mills were inherited by his grandson by marriage, Alexander Pirie, who founded the well known firm of Alexander Pirie & Sons. Smith's nephew, Charles, who had gained experience in paper making from his uncle at Stoneywood, now withdrew to found another paper mill nearby.[4]

His partner was Charles Davidson, the founder of the firm which bears his name. He was the son of an Aberdeenshire dyer and had been trained as a millwright in Aberdeen, working in his youth at Grandholm Waulk Mill on the opposite side of the river Don to Stoneywood. The partnership with Charles Smith lasted for only a few years, however, and by 1811 he had returned to Grandholm Mill.[5]

He set up in business on his own account in 1811 by obtaining from James Forbes of Seaton a 57 year lease of land at Mugiemoss together with the dwelling house, barn and byre for an annual rent of £18. The land included an old mill lade originally 8–10 feet wide, choked with sand and rubbish, which Davidson proposed to clear out to bring water to power the mills he was about to erect.[6]

Two further agreements consolidated his position at Mugiemoss. In 1826 the lease was extended by 19 years to 1886 and now included land previously rented by John Mackie with an adjacent strip belonging to the landlord, the rent for the whole of the lands and buildings being increased to £63 a year until 1867 and £97 thereafter. Finally in 1833, when he was described in the document as a papermaker, a new lease was drawn up giving him additional land, the mansion house of Mugiemoss itself, and fishing rights for salmon and trout along an adjacent stretch of the river Don. The lease for all this land and property was extended to run for 114 years expiring at Martinmas (11 November) 1947, at a rent of £75 a year for the first 34 years and £110 a year for the remainder of the term.[7]

At various times between 1811 and 1830 partly utilising his earlier experience of textiles manufacture the mills erected at Mugiemoss fulled woollen cloth, beat flax to prepare it for flax spinning and linen manufacture, and ground fermented tobacco leaf into snuff to cater for the habit then fashionable among the upper classes. Paper making came later, probably beginning in 1821, and soon displaced these other activities. It is significant that the 1833 lease of Mugiemoss described Charles Davidson simply as a papermaker and a year earlier the first surviving national list of papermaking firms in Scotland recorded him as a manufacturer of brown and grey paper possessing one of the 32 new Papermaking machines installed in the industry over the previous 20 years to replace the old hand made process.[8]

Nevertheless although Charles Davidson had established an independent business on a firm footing in the paper industry the business remained on a relatively modest scale when he died in 1843. His personal estate was valued at £942, including a half share in the firm, and the business itself was estimated to be worth £1,740.[9] The following year the *New Statistical Account of Scotland* noted only a single Papermaking machine at the mill and although the size of the labour force was not directly indicated it was probably less than one hundred persons, for all four paper making firms in the Aberdeen area including the largest – Alexander Pirie & Sons with two Papermaking machines and 150 employees – had a combined labour force of no more than 300–400 persons.[10]

THE SECOND GENERATION

William and George Davidson, who had assisted their father in running the business for several years before his death, now took charge and gradually enlarged the scale of operations.[11] Little is known of these early years but by the beginning of 1852, when the earliest surviving record in the firm's archives commences – a Private Ledger covering the years 1852–65 – the buildings at Mugiemoss were valued at £3,017 and the machinery at £4,833, giving a combined total of £7,850, exclusive of stocks of materials and finished products worth £3,071.[12]

Further dreams of expansion received a rude shock when on 29th June 1853 a large scale fire began in the main three storey building at the works where pulp was produced for papermaking. The alarm was sounded about 11.15 p.m. but fanned by a brisk breeze the whole building was soon ablaze and after an hour the roof collapsed followed by the wooden floors, leaving the building in ruins. No lives were lost and the firefighters were able to prevent the fire spreading to other buildings on the site. Nevertheless, damage to machinery and the buildings amounted to some £3,000 – of which nearly £2,200 was recovered from the Northern

Assurance Company, together with another £700 for raw materials destroyed. However, the most severe effect of the fire was the loss of future production and it was clear that several months would elapse before the Mugiemoss Works were again running at anything near full capacity.[13]

Faced with this depressing situation it was felt that bold measures were required and instead of simply rebuilding, the opportunity was taken of enlarging the Works substantially. The project was made possible by obtaining a substantial loan from the North of Scotland Bank which was secured by a lease on the firm's land, buildings and machinery.[14] The New Mill Account which appeared as a fresh item in the Private Ledger, sadly records no details of the expenditure involved but £14,000 had been spent on the new mill by 31st December 1855 and a further £2,000 was invested in the next two years.[15]

It was also decided to obtain a lease on the adjacent Waterton Mill formerly used for papermaking by Thomas Jaffray.[16] Hence by the late 1850s the firm possessed three papermaking machines and the potential for a much greater output.

These changes were followed by a major diversification in output – the production of paper bags. After the middle of the 19th century retail customers of paper firms producing various kinds of wrapping, packing, and grocery papers were increasingly using some of these papers to make up paper bags in which they sold their goods to the individual consumer. The bags would be made by hand, as for example on the counter of a grocer's shop, or by simple hand operated machinery. However a few paper manufacturers and wholesale dealers in paper explored the possibility of supplying retailers with ready made paper bags in various sizes made by power driven machinery. The appeal of cheapness and convenience for the retailer might be reinforced by printing on the bag his name and address together with some information about his principal wares.[17]

William Davidson's second son George, who had an inventive mind, was one of these pioneers. He took out patents in December 1859 and June 1863 for a machine operated by power to make paper bags either from a roll of paper or from the web of paper coming directly from the dry end of the papermaking machine. When perfected the bag machine was capable of producing some 4,000 paper bags per hour.[18]

This machinery and the installation of hand operated machines for making paper bags of kinds and sizes for which efficient power driven machinery was not yet devised, began to transform the character of the firm's output. A glimpse of this can be seen from the firm's Private Ledger. Before 1859 scarcely any bags were recorded at the annual stocktaking, although stocks of paper held regularly exceeded £1,000 in value, yet by July 1863 paper bags of all kinds accounted for more than a quarter of all paper stocks out of a total of more than £3,000, whilst the stock of bags increased not only in the range of sizes and qualities but also in variety, to include bags for seeds, soda, flour, draper's goods, tea, and biscuits.[19]

The growth of output and diversification into paper bag production was accompanied by a more conscious effort to attract distant customers. Stocks of paper were being held in London by the beginning of the 1850s and the need for more permanent arrangements to serve the large and wealthy London market with its varied demands was secured in September 1858 by the 57 year lease of 80, Upper Thames Street, which was conveniently situated in the heart of the city, a few yards from the Thames, for an annual rent of £200.[20]

More tentative and more ambitious were the steps taken to secure customers in the future Dominions. In 1858 and 1859 the Private Ledger records both the overall costs and receipts resulting from 'adventures' to Quebec and Montreal in Canada, whilst in 1863 there was a note of a similar 'adventure' to Melbourne.[21]

The fruits of all this energy and enterprise can be seen in the growing volume of business recorded in the Private Ledger. The increase in sales was masked partly by the excise duty which was levied on all paper produced in the United Kingdom until October 1861, when the duty was abolished.[22] The duty artificially inflated the price of paper in the earlier years shown in Table 1. However, if the sales figures are arbitrarily adjusted by deducting the sum paid in excise duty each year the figures in the final column of Table 1 give a reasonable indication of the growth of the firm's sales from 1852 to 1865.

Sales in 1852 and 1853, before the serious fire at the end of June 1853 and the erection of the new mill, were running at some £9,000 a year. In 1856 and 1857 after the completion of the mill and the installation of additional papermaking capacity, sales averaged more than £18,000 a year. By the early 1860s sales had increased still further exceeding £27,000 in each of the three years

Table 1: Income, 1852–1865

Period Covered	Paper Sales £	Other Income £	Total £	Paper Duty £	Paper Sales Adjusted Total[1] £
1 January 1852–30 June 1852	8,818	11	8,829	3,063	5,755
1 July 1852–30 June 1853	13,147	19	13,166	5,778	7,369
1 July 1853–31 December 1855	29,993	79	30,072	10,379	19,614
1 January 1856–31 December 1856	27,155	180	27,335	10,393	16,762
1 January 1857–31 December 1857	30,062	99	30,161	10,418	19,644
1 January 1858–31 July 1858	18,202	151	18,353	6,919	11,283
1 August 1858–31 January 1859	16,508	128	16,636	6,407	10,101
1 February 1859–31 July 1859	16,818	108	16,926	6,292	10,526
1 August 1859–31 January 1860	15,902	96	15,998	5,339	10,563
1 February 1860–31 July 1860	19,002	113	19,115	5,700	13,302
1 August 1860–31 January 1861	18,591	65	18,656	5,876	12,715
1 February 1861–31 July 1861	19,273	56	19,329	6,051	13,222
1 August 1861–31 January 1862	16,755	160	16,915	2,163	14,592
1 February 1862–31 July 1862	13,827	85	13,912	—	13,827
1 August 1862–31 January 1863	13,933	24	13,957	—	13,933
1 February 1863–31 July 1863	13,697	86	13,783	—	13,697
1 August 1863–31 January 1864	15,173	61	15,234	—	15,173
1 February 1864–31 July 1864	15,259	60	15,319	—	15,259
1 August 1864–31 January 1865	12,009	20	12,029	—	12,009
1 February 1865–31 July 1865	13,798	87	13,885	—	13,798

Source: Private Ledger, 1852–65
Notes: 1 It has been assumed that the whole of the excise duty was passed on to customers in higher prices and the adjusted figures in column 5 were obtained by deducting column 4 from column 1.

from 1 February 1862 to 31st January 1865. Overall the increase represented nearly a three fold expansion in sales since 1852.

The growth of profits, summarised in Table 2, was less impressive. Indeed the final investment in the new mill scarcely had been completed when the effects of the severe recession of 1858, felt throughout the United Kingdom, threatened financial disaster. Several large customers were unable to pay their debts and bad debts reached the unprecedented sum of £877 in London, £191 in Edinburgh and lesser sums elsewhere. There was a further loss of £545 on government contracts, a loss of £124 on trade with Quebec, and overall the deficit on the six months' operations ending on 31st July 1858 amounted to £2,803.[23]

The North of Scotland Bank not unnaturally was now concerned about the firm's financial future and in particular the prospects of securing the eventual repayment of its large loan. Robert Fletcher was therefore sent to make a thorough investigation of the firm's accounts and to report to the manager of the bank in Aberdeen. Thus began a series of regular six monthly reports on the firm's finances of which, as we shall see shortly, a continuous series covering the years 1865 to 1875 has survived in the University archives. Fletcher recommended some improvement in the firm's accounting procedure but the basically sound nature of the business was reflected in the rapid recovery in profits.[24]

There was a small loss of £57 over the next six months but thereafter over the six and a half years to 31st July 1865 a respectable level of profits was achieved. Profits exceeded £500 in each of these years, averaging £1,049, and the overall performance was marred only by a loss of £453 in the half year ending 31st January 1865 when a whole month's production was lost whilst a new steam engine was installed. Moreover more than £10,000 was spent on repairs and renewals of buildings and machinery over these years exclusive of a further £2,900 found out of profits for investment in the new steam engine and other additional plant.

The regular six monthly reports to the Manager of the North of Scotland Bank which have survived from 1865 onwards permit a closer scrutiny of the firm's progress over the next ten years. The picture given by the reports can be supplemented by data from a Minute Book of Meetings held every six months by William Davidson's sons to discuss the accounts and plan future policy after their father became mentally incapable of running the business in

Table 2: Profits,[1] 1852–1865

	Profit £	Loss £
1 January 1852–30 June 1852	1,611	—
1 July 1852–30 June 1853	107[2]	—
1 July 1853–31 December 1855	446	—
1 January 1856–31 December 1856	534	—
1 January 1857–31 December 1857	205	—
1 January 1858–31 July 1858	—	2,803[3]
1 August 1858–31 January 1859	—	57
1 February 1859–31 July 1859	506	—
1 August 1859–31 January 1860	21	—
1 February 1860–31 July 1860	607	—
1 August 1860–31 January 1861	527	—
1 February 1861–31 July 1861	48	—
1 August 1861–31 January 1862	633	—
1 February 1862–31 July 1862	531	—
1 August 1862–31 January 1863	607	—
1 February 1863–31 July 1863	427	—
1 August 1863–31 January 1864	427	—
1 February 1864–31 July 1864	1,324	—
1 August 1864–31 January 1865	—	453[4]
1 February 1865–31 July 1865	127[5]	—

Source: Private Ledger, 1852–1865

Notes: 1 Surplus available, after meeting all expenses (including repairs), for depreciation, investment, placing in reserves and distribution. The expenses include sums drawn by William Davidson; the maximum recorded in any one account was £376.

2 The profit figure was reduced to this level as the result of damage caused by fire in June 1853. The bulk of the damage was covered by insurance but an additional £810 for damage to machinery and buildings had to be found from the firm's own funds.

3 The size of the deficit can be explained partly by the high level of bad debts, more than £877 in London and £191 in Edinburgh, which reflected the severe recession in the economy in 1858. There was also a loss of £545 on government contracts.

4 A note in the accounts attributed the poor results partly to the loss of about 30 days' production 'in putting in new work in connection with the new steam engine'.

5 The continued poor results were attributed mainly to the breakdown of the new steam engine, which provided only 20 days of steam power in the half year.

Table 3: Output of paper, 1858–1874

Financial Year Ending 31st July	Output (Tons)
1858–59	896
1859–60	938
1860–61	991
1861–62	1,010
1862–63	1,003
1863–64	1,066
1864–65	905
1865–66	1,105
1866–67	1,409
1867–68	1,571
1868–69	1,730
1869–70	1,775
1870–71	2,159
1871–72	2,849
1872–73	3,636
1873–74	3,482

Source: AUA, MS 2769/I/30/5

1870. It is apparent from Figure 1 that paper production which had hovered between 900 and 1,000 tons a year from 1858 to 1865 increased markedly after this period. By 1867–68 output exceeded 1,500 tons, the 2,100 tons mark was passed in 1870–71 and by 1872–74 production was running at more than 3,400 tons a year. Overall there had been a more than three fold expansion in the course of nine years.[25]

The increase in output was made possible by a sustained and substantial investment programme. In the five years 1866–70 a total of £4,262 was invested in additional plant, an average of £852 per year.

Over the next four and a half years to 31st January 1875 these efforts were far surpassed. Expenditure on new plant regularly exceeded £2,000 a year and in all, nearly £15,000 was invested.

One object of the investment was to reduce the annual bill for repairs and renewals to the existing buildings and equipment. Nevertheless more than £22,000 was spent on repairs and renewals over the nine and half years in addition to investment in new plant.

The productive capacity of No 1 and No 2 Papermaking Machines was increased by various means including the installation of extra cylinders and the provision of new steam for additional motive power. A new chimney stalk was erected and new rag boilers and rag engines were purchased to process more raw materials. The output of finished products was also boosted by the erection of a new Finishing House designed to house a new steam engine and cutter, a ripping machine and additional bag making machines.[26]

However, increased production was not the sole objective. A number of projects were designed specifically to economise on fuel consumption, spurred on by the rising price of coal. The main Water Wheel was completely renewed and extensive improvements made to the machinery connected to it, the tail race was widened and deepened and the lade or channel leading to the water wheel improved. At the same time a fourth steam boiler costing £450, using Root's patent, and a new hot water pump were installed to make more effective use of steam power.[27]

Improving the quality of the firm's products also exercised considerable attention. Efforts were made to minimise irregularities in the cleanness and thickness of the papers produced and this policy was re-affirmed in February 1873 when it was resolved 'that our utmost efforts be given at this time to keep up the quality and regularity of our various papers so as to establish our name and trade firmly'. New stuff chests and stuff mixers were installed to ensure that the pulp delivered to the papermaking machines was of better quality whilst orders were placed for a Dusting and Cleaning House with dusting machinery to improve the fibrous raw materials used to make the pulp.[28]

The overall effectiveness of the various measures adopted can be measured in financial terms by the profits generated from 1865 to 1875. It can be seen from Table 3 that there was a surplus on the firm's operations in each of the six monthly accounts over the nine and a half years to 31st January 1875. The level of profits never fell below £1,500 in a year, rising to a maximum of £10,777 in 1873. Overall the upward trend was quite clear, profits averaging some £2,900 annually in the five years 1866–1870 and more than £7,400 in the next four and a half years to January 1875.

Moreover, even when the substantial programme of investment in new plant had been met there was still a healthy surplus available for all other purposes. The principal concern in fact was the

Table 4: Income, Profits and Investment, 1866–1875

Half Year Ending	Income £	Expenditure £	Surplus or Deficit £	Investment in Additional plant £	Profits remaining £
31 Jan 1866	18,670	18,018	652	166	486
31 Jul 1866	19,439	18,527	911	173	739
31 Jan 1867	22,833	22,233	600	613	–13
31 Jul 1867	25,957	24,299	1,658	995	664
31 Jan 1868	24,639 [1]	23,551	1,088	87	1,001
31 Jul 1868	25,531	22,725	2,805	672	2,133
31 Jan 1869	25,623	25,052	571	153	419
31 Jul 1869	28,009 [2]	24,210	3,799	619	3,180
31 Jan 1870	25,194	24,167	927	222	705
31 Jul 1870	28,056	26,492	1,564	562	1,002
31 Jan 1871	27,562	27,263	199	2,098	–1,899
31 Jul 1871	31,440	29,836	1,603	876	728
31 Jan 1872	32,747	31,240	1,507	1,146	361
31 Jul 1872	40,073	34,625	5,449	1,140	4,309
31 Jan 1873	47,745	42,000	5,745	2,426	3,318
31 Jul 1873	51,261	46,229	5,032	2,382	2,650
31 Jan 1874	48,907	45,570	3,336	869	2,468
31 Jul 1874	43,963	40,868	3,096	1,265	1,831
31 Jan 1875	51,436	43,783	7,653	2,635	5,018

Source: See Figure 1

Notes: 1 Includes £854 for loss of profit occasioned by a fire in October 1867, debited to the amount recovered from insurance.

2 Includes £1,344 profit on the sale of part of the property at Regent Quay, Aberdeen Harbour.

reduction of the debt owed to the North of Scotland Bank. William Davidson and his sons had decided by January 1868 to repay at least £1,000 a year and by agreement with the Bank the six monthly accounts from 31st January 1868 onwards made the position quite clear by separating the firm's current account with the bank from an account recording the amount of the loan outstanding. The latter then stood at £24,000 and by 31st January 1875 this had been reduced to £16,000. An additional loan granted a few months earlier specifically to develop a new sales department amounted to £3,476 but the surplus in the current account was then £12,744 thus reducing the real debt to the bank to £6,732.[29]

Throughout most of this period until his death in March 1873 William Davidson was the dominant figure in the firm and he was the sole owner when he died. He left behind him a firm which was more valuable and far stronger than he had inherited in 1843. The detailed inventory of his possessions drawn up after his death valued his personal estate and effects in the United Kingdom at more than £32,000. Of this sum the stock in trade, moveable machinery and the balance of debts owing to the business in Scotland represented more than £19,100 whilst the London side of the business added a further £11,300 making an overall value for the firm – exclusive of permanent buildings and fixtures – in excess of £30,000. This may be compared with the valuation of £1,740 placed on the business thirty years earlier.[30]

Data from the firm's accounts available over the shorter period from 1852 onwards and covering fixed assets as well as the other assets suggests that some of the growth had already occurred by 1852, when the firm's assets including buildings worth £3,017 were valued at nearly £11,000. However, over the next 21 years the combined value of the fixed and moveable assets and debts owed to the firm increased almost five fold to more than £51,000.[31]

The picture of the business in 1873 can be supplemented by data given a few months earlier to the Royal Commission investigating the pollution of rivers in Scotland. The three papermaking machines and other machinery were powered by both water and steam yielding approximately 200 nominal Horsepower. Some 4,000 tons of coal, 3,000 tons of rags and more than 600 tons of chemicals were consumed annually and 216 persons were employed at Mugiemoss.[32]

Paper bag production was not mentioned specifically, but the firm retained its position as one of the pioneers in the field and the *Paper Mills Directory* in 1871, for example, recorded only four other paper manufacturers who produced paper bags, three of them in England.[33]

CONVERSION TO A LIMITED COMPANY

The prosperous firm built up during William Davidson's lifetime had not been created without assistance from other members of the family. His brother George had left in the 1850s to pursue an independent career but William's five sons in turn, commencing with the eldest, Charles, who reached the age of 21 in 1853, joined their father in the business. The sons made a valuable contribution to the growth of the firm, particularly George, who patented the paper bag making machinery, and Charles, who took charge of the London office until his death in April 1870.[34] Nevertheless they remained salaried employees without a stake in the firm, a situation which they found increasingly irksome as they married and started their own families. On several occasions the desirability of taking the sons into partnership had been discussed but no action had been taken, no doubt because the existing situation suited their father, but ostensibly on the grounds that the North of Scotland Bank's lease on the land, buildings and machinery of the firm, held as security for the large loan granted some years earlier, created a legal obstacle.[35]

However in William Davidson's last years the balance of power changed dramatically. On 31st July 1869 he was persuaded to sign an agreement designed to place the operations of the firm on a formal basis. This placed his sons in a much stronger position, giving each one a share in the firm's surplus profits and accepting the principle of majority decisions, albeit with a casting vote for their father, together with the right to convert the firm into a partnership in the future if the majority of sons wished it.[36]

Some fifteen months later William Davidson suffered a stroke. He recovered physically but with his mental faculties impaired and the sons took over the running of the firm.[37] Yet tantalisingly because the agreement of 1869 had not been signed before independent witnesses, which would have made it without question legally binding, and perhaps also in deference to their father's state of health and his feelings, no steps were taken at this time to reconstruct the firm as a formal partnership.

The ambiguous legal status however made the sons decide by August 1872 to take legal advice about various financial aspects of operating the firm including the possibility of forming a partnership under the 1869 Agreement. The opinion of counsel in Edinburgh was sought eventually and at the beginning of March 1873 the sons were advised that they might achieve their objectives by applying to the legal authorities to appoint a *Curator Bonis* who would take charge of William Davidson's financial affairs and by recommending for the post someone in whom they had confidence and who had their interests at heart. The procedure was never tested, however, for in a final ironic twist to the tale, their father died a mere sixteen days later.[38]

In the new circumstances, and with a growing confidence generated by the increasing profitability of the enterprise, the sons began to think about a rather more novel type of organisation for the firm – conversion to a limited liability company. This form of business organisation was becoming more popular in Scotland by the early 1870s. It offered the distinct advantage of restricting the financial liability of businessmen to the loss of their shares in the firm rather than suffering financial ruin if their firm collapsed, whilst in the more likely circumstances of continuing profitable expansion contemplated by the Davidson brothers it would be easier at some future date to raise additional capital and to dispose of some of their share in the business by selling shares to the investing public when extra cash was required for exceptional personal and domestic needs.[39]

Once conversion to limited company status had been decided upon, a number of important details had to be settled. The principal one was the value to be placed upon their business. The figure placed on the value of the firm's assets in the half yearly accounts provided a neat and convenient answer but it was felt that a broader approach might be more realistic. Therefore an internal memorandum on the whole subject was prepared. It began by observing:

1 The value of the business must be estimated as that of a going concern, which it is.
2 The value would therefore include not only all book debts (less liabilities), all stocks at market prices, all machinery and plant

at working value but also many things of a much less tangible form such as Goodwill of the business, probable value of several leases running and so forth.

3 The very greatest difficulty of opinion might exist not only as to these less tangible items but also as to the value of machinery and plant of all kinds, and hence a satisfactory valuation of the whole *taken piecemeal* could never be arrived at.

Hence it proposed that the basis of calculation should be the rate of return obtained from capital employed in the business. The net profits recorded in the eight previous accounts, each covering half a year, after making allowance for the depreciation of the plant, amounted to £22,402, giving a yearly average of £5,600. If 10 per cent was taken as a reasonable return – and the author believed that this figure would be generally expected by investors, especially if shares offered for sale on the open market were to reach par value – then the capitalisation of the new company should be £56,000.[40]

In fact, when the company was incorporated on 18th February 1875 the authorised capital was £80,000 divided into 8,000 shares of £10 and the agreement for the sale of the existing business to the new company valued the firm at £60,000. Of this sum £35,000 represented payment for tools, moveable machinery, stock in trade and raw materials, £5,000 was to be paid for the Goodwill of the business, and the remaining £20,000 represented the value placed on the various leases held by the firm. Payment was to take the form of £21,000 in cash, plus 975 fully paid up shares of £10 for each of the four brothers, who became the first directors of the company.[41]

In essence therefore the new form of organisation, as for a large majority of family firms adopting limited company status in the second half of the nineteenth century, made virtually no initial difference to the running of the firm. Control and the bulk of ownership remained in the hands of the Davidson brothers. No other directors were appointed for more than 20 years and although the company was permitted to issue 8,000 shares, which would raise an extra £20,000 above the purchase price, it was more than 10 years before the issued capital exceeded £60,000.

Chapter 2

Expansion And Prosperity, 1875–1900

NEW PRODUCTS AND NEW MARKETS

During the final quarter of the nineteenth century the firm continued its sustained growth. The expansion was fed by a steady stream of investment in addition to substantial sums expended each year on repairs and renewals to existing equipment.

The most spectacular project was the acquisition of the adjacent Bucksburn Mills in 1876. The leasehold property and plant were purchased for £5,321 from James Kilgour an Aberdeen draper who had acquired it after the business failure of James and Alexander Howie, woollen manufacturers at the mills.[1]

The main motive for the purchase was the provision of additional water power and the Davidson brothers, alarmed by a steep rise in coal prices, had resolved in February 1873

> That if it is found practicable, overtures be made for the purchase of Buxburn Woolen Mills for the sake of the water power, and that this be seen about without delay.

However the decision was also influenced by uncertainty over the renewal of the lease of Waterton Mill which expired in 1880. The new landlords, Alexander Pirie & Sons, who owned the adjacent Waterton House, objected to the smoke from the mill spoiling their amenities. The lease was in fact not renewed and subsequently the mill was demolished.[2]

Few details of the other items of investment have survived. The directors' annual report to shareholders at this period, for example, did not elaborate on the investment total given in the accompanying accounts, and only one Directors' Minute Book, covering the years 1898–1900, has survived, giving information on the purchase of some items of equipment. However although it is impossible to break down the totals into different categories the overall impression from the data summarised in Table 5 is quite clear. More than £60,000 was invested in new equipment and buildings between 18 February 1875 and 31 July 1899. Of this sum £10,725 was invested in the first four and a half years, an average of nearly £2,400 a year, whilst in the 1880s more than £23,000 and in the 1890s nearly £27,000 were invested, averaging at least £2,300 a year in both these decades.

Some of the investment was committed to the transformation of

Table 5: Investment, 1875–1899

Year Ending 31st July	Investment in New Plant[2]
1875[1]	3,092
1876	3,142
1877	1,867
1878	1,375
1879	1,249
1880	1,823
1881	1,555
1882	2,858
1883	2,534
1884	1,151
1885	3,435
1886	1,399
1887	1,354
1888	4,387
1889	2,619
1890	2,067
1891	2,854
1892	5,329
1893	2,255
1894	1,653
1895	493
1896	2,339
1897	5,880
1898	2,379
1899	1,643

Source: *Annual Report and Accounts*, 1875–99
Notes: 1 Half Year Ending 31 July.
2 Includes the London Warehouse as well as Bucksburn, Mugiemoss, and Waterton (up to 1880).

the firm's sales network. As the volume of sales increased the existing arrangements for the distribution of its products – based on Mugiemoss and the Aberdeen and London warehouses – came to be seen as insufficient so that a network of provincial offices, each with its own warehouse and sales staff, was created. The premises were often rented on relatively short leases, which economised on capital input and made it easy to switch to a different location when necessary.

The first of these provincial warehouses was located at 51 St Enoch Square, Glasgow, leased in July 1882 from the Glasgow and South Western Railway Company for five years at a rent of £60 for the first two years and £70 thereafter. However in September 1887 the railway company granted a lease over other premises in the square numbered 59–63 for five years at £140 a year and this lease was subsequently extended with an increase in rent to £165.[3]

An office at Newcastle-upon-Tyne was established soon after this. Number 34 Dean Street was leased in November 1883 for 10 years at £70 a year, but the accommodation proved inadequate for in November 1886, much larger premises were rented on the High Bridge for £80 a year, rising after the second year to £95, and this lease was later extended.[4]

Liverpool in 1888 and Leeds in 1892 followed. However the Leeds office was closed down after a further year and instead the network of bases was completed in April 1894 by leasing a warehouse in Castle Terrace, Edinburgh from the Local School Board for five years at a rent of £65 a year, the lease being renewed at a later date. Customers in the whole of the South East of Scotland were better served from the Edinburgh office whilst the office in Newcastle-upon-Tyne could readily seek out and supply customers in the Yorkshire area.[5]

Meanwhile in London, the oldest sales office outside Aberdeen and the most important, a decision was made to seek new premises, even though the lease on 80 Upper Thames Street did not expire until 1915. In March 1884 the lease on premises at 119 Queen Victoria Street, which had been rented by C.J. Ingram for 21 years at £450 a year expiring in 1895, was assigned to Davidsons for the sum of £775. However this move proved only a temporary solution to the firm's needs. In November 1894 Paul's Pier Wharf and 23 Upper Thames Street, situated a few yards to the west of the original site, were leased from the Earl of Radnor for 62 years at an annual rent of £1,044. The new premises offered space for future expansion, whilst the high costs were partially offset by a policy of sub-leasing some space to other tenants.[6]

A survey of salaried employees in the 1890s, summarised in Table 6, throws light on the completed sales network. As can be seen Newcastle, Liverpool and Edinburgh each had a manager and three commercial travellers with supporting staff of clerks and warehousemen, plus in Newcastle a porter and an office boy, whilst Glasgow boasted an additional traveller, a cashier, two clerks and two porters. The combined sales staff of the four branches thus consisted of 36 persons of whom 13 were specifically described as travellers.

The London office during the 1890s was the base for four commercial travellers with two clerks in support. However its greater importance was emphasised by the presence of a second manager and by a managing director, Alexander Davidson. He had taken charge of the London operations at the original site in Upper Thames Street after the death of his brother Charles in 1870 and his decision to reside permanently in London as the senior Davidson director reflected the Board's determination to exploit the London market. The Company Secretary, replying in 1898 to a disgruntled shareholder, who complained among other matters about the extra expense which would be incurred if additional

Table 6: Salaried Staff Employed at the Company's Warehouses in the 1890s[1]

	Number Employed					
Warehouse	Manager	Clerk	Traveller	Warehouseman	Others	Total
Edinburgh	1	2	3	1	—	7
Glasgow	1	2	4	2	4	13
Liverpool	1	2	3	2	—	8
London	3[2]	2	4	—	—	9[3]
Newcastle	1	1	3	1	2	8

Source: Lists of Salaried Staff prepared for Income Tax Returns

Notes: 1 Data on the Aberdeen warehouse is not available. The London figures are for 1893 and 1894, the other figures for 1897.

2 Includes Alexander Davidson a director.

3 Data on the staff of the separate Printing Plant is not available.

London based directors were appointed since they would have to travel to Board meetings in Aberdeen, observed that a large part of the company's business was in London. The Directors' Minute Book for 1898–1900 fully confirms the truth of this observation for in 1898 and 1899 sales of the company's products from the London Office exceeded those from all the other warehouses combined, amounting to 50 per cent of sales in 1898 and 52 per cent the following year.[7]

Export markets were not neglected in the search for additional customers. Some orders were obtained rather passively from merchants or perhaps individual customers overseas, for as late as 1886 the firm was advertising that goods for export were 'Press Packed at the mills at cost price and delivered at any of the docks in London, Liverpool and Glasgow.' But agents under contract in Australia and Scandinavia were acting specifically for C. Davidson & Sons, selling their felt papers, before the end of the century.[8]

Moreover the growth of business in Australia led to a further initiative. The pioneering visit to Melbourne in 1863 had been followed by further contacts, including 'an adventure' to Geelong on the adjacent coast which had incurred bad debts of nearly £100 between 1st August 1873 and 31 January 1875.[9] Perseverance with such initiatives led to the growth of a regular and substantial Australian business but one centered mainly in Sydney and New South Wales. By 1893 the business was sufficiently well established to justify the formation of a separate company to take over the trade in New South Wales. The Colonial Paper Company Limited was incorporated in England with an authorised capital of £10,000. Its headquarters was the London office of Davidsons and Alexander Davidson and John Mackie, a manager at the same office, were directors. The new company was to be supplied with paper bag machinery and other equipment in Sydney capable of converting 25 tons of paper into bags per month and Davidsons also undertook not to carry on a separate trade as dealers or makers of paper bags anywhere in New South Wales.[10]

Investment was also directed by the Davidson directors towards increasing the range of products available to customers. The *Paper Mills Directory*, which listed all the papermaking firms in the United Kingdom together with their products, recorded 10 different types of paper made by C. Davidson & Sons in 1876 and 16 types in 1890, including 8 not listed in the earlier year.[11]

The firm continued to produce its traditional brown packing and wrapping papers such as *small hands*, named from the size of the paper, *skips* for lining crates and packing cases, *mill wrappers*, and *cartridges*, tough papers for covering pamphlets, catalogues and cheap books or for use as drawing paper. The range of products in 1876 was completed by paper bags of various kinds, papers for the grocery trade, and *middles* – common cardboard made into bus and tram tickets or used for *pasteboard*, high quality surfaced cards constructed by pasting good quality paper onto both sides of the inferior *middle*.

The additional products on offer by 1890 included *manilla papers*, which were tough, durable, and flexible and were used for parcel labels and book binding. *Butter papers*, which were transparent and greaseproof, were made for wrapping butter, lard, and other greasy substances, whilst *duplex papers* offered the customer two differently coloured surfaces, sought after for use in advertisements and programmes for the theatre or other events.[12]

There was also increased emphasis on making papers with a glazed surface. This was produced in a separate process after the paper had left the dry end of the papermaking machine by passing the paper through a super calender which consisted of a series of alternate rolls of polished steel and compressed paper. However if an even more polished surface was desired a friction calender might be employed, consisting of large compressed paper rolls and small steel ones, which revolved at a higher speed, thus imparting considerable friction to the sheets of paper passing through. The glazing process reduced the thickness of the paper whilst increasing its tensile strength and imparted a polished surface with a cleaner appearance which was not only lighter and more attractive but also more repellent to dirt and water.[13]

However perhaps the most significant introduction, leading the firm towards quite different retail customers, was the production of felt papers for various purposes. Thick soft felted papers, generally made from waste papers, were sold as carpet felt for the purpose of covering floorboards so that they would not mark the carpet laid on top and would give a better feel to the floor covering. On the other hand strong thick paper coated with tar was sold for use as roofing felt or for laying on stone floors underneath linoleum, which was thus protected from damp and from the abrasive effect of an uneven surface.[14]

In order to exploit the market for felt products more fully it was decided to establish a separate limited company, the Boulinikon Felt Company, which was incorporated with an authorised capital of £10,000 in February 1883. The registered office was Paternoster Square in the city of London and the four directors, who held the bulk of the shares, were Alexander, John and David Davidson together with John Mackie, who was then an accountant employed in their London office.[15]

In 1884 a number of technical improvements were made to their paper felts. Alexander Davidson took out a patent for coating the surface of waterproof felt with a powder made by grinding the husks of oats, barley, wheat or similar grains, although sawdust might be used as an alternative. The powder was applied to the surface of the felt which had just been covered with hot tar to make it waterproof and became embedded in it, thus solving the problem commonly encountered that waterproof felts, even when the tar coating had cooled and set, were dirty to handle and tended to stick together when rolled or laid upon each other.[16]

More than a decade later he was still seeking improvements and by chance a series of letters have survived from 1895 showing him making enquiries about the costs of manufacturing roofing felts in three different colours. In the course of this he obtained information from Denmark about a patent centrifugal tar separator used at Copenhagen Gasworks which produced better quality water free tar, and it was estimated by the Danish supplier that the installation of such a separator at Mugiemoss would prove more economic in making waterproof paper and roofing felt than buying pure tar from outside suppliers.[17]

However the most famous innovation, which became well known to customers purchasing carpets in the United Kingdom, was the production of Cedar Felt. In an era when synthetic carpet fibres and underlays were unknown and safe chemical protection was over the horizon, the addition of cedar dust to carpet felt provided a deterrent to moths and other insects as well as providing a pleasant aroma. Alexander Davidson took out a patent in 1884 to make carpet felt, incorporating aromatic woods such as cedar, sandalwood or pine. Dust made by grinding the wood, or an essence obtained by boiling down or distilling the wood, was added to the pulp before it went to the Papermaking machine. However it was found that the cedar dust produced a felt speckled with particles of wood whilst the addition of too much dust unduly weakened the felt. Neither problem was insuperable but in 1898 he patented an improved method of manufacture whereby the cedar dust was sprinkled on the top of a layer of pulp as it was carried along on the travelling wire cloth in the Papermaking machine, and this layer was then covered by another layer of pulp so that the cedar dust, which could be used in a much greater concentration than with the earlier technique, formed a sandwich between two layers of paper.[18]

The improvement in products was reflected in rising sales. Paper felt products were sold in outlets throughout the United Kingdom and by the early years of the 20th century some major furnishing shops and department stores were regular customers, receiving special discounts on orders. The 38 shops listed in London included familiar names such as Harrod's, Maple & Co. Ltd. of Tottenham Court Road, and Peter Jones of Chelsea, whilst elsewhere Rylands & Sons Ltd of Manchester, Ray & Miles of Liverpool, Keen & Scott of Birmingham, Arnott & Co. Ltd. of Dublin, and six shops in Glasgow including Wylie & Lochhead, were given similar favourable terms.[19]

The mainstay of manufacturing operations for C. Davidson & Sons, nevertheless, remained the production of paper bags though here too the product was improved and the variety widened. The protection afforded by the original paper bag patents taken out by George Davidson jnr. expired soon after registration as a limited company but the directors were determined to remain abreast of further technical advances in this field, both by their own inventive efforts and by obtaining licences to use machinery devised by other inventors.

An agreement was signed with Job Duerden & Co. paper bag machine makers of Burnley in 1887 for the supply of the firm's No.1 Machine to make Block Bottom Bags, utilising a patent developed by an American engineer, Charles O'Steans of Boston, which Duerden was permitted to employ in the U.K. The machine made bags of various lengths from 5.5 to 9.25 inches and widths from 4 to 7 inches and could use paper of any thickness from 15 pounds to 65 pounds weight in Double Crown size. The price was £300, exclusive of power belting and driving which Davidsons supplied, and Duerden promised to supply up to another seven machines at the same price within two years from the delivery of the first one, with the proviso that none of the machines could be sold or used outside the United Kingdom.[20]

Further American paper bag patents were obtained in 1898 by means of an agreement with the Anglo American Self Opening Square Paper Bag Machine and Manufacturing Company Limited, which had purchased, soon after its formation in April 1888, the patent rights of the International Paper Bag Machine Company of New York. C. Davidson & Sons agreed to purchase the goodwill of the business together with trade marks, designs, certain patents, the stock of paper and paper bags and the loose plant and machinery for the sum of £1,773, part of which was to be paid by allocating 1,000 of Davidson's unissued £1 shares to the two directors of the Anglo American Self Opening Square Paper Bag Machine and Manufacturing Company, Henry Fricker and John Thomas Bibby, who were both living in London.[21]

Meanwhile two improvements in machinery to make block bottomed and square bottomed paper bags had been patented by the firm. The first patent was taken out in 1893 by Alexander Davidson, the director in charge of the London Office, and the second two years later by John Davidson and William Smith Murray who were based at Mugiemoss.[22]

Other bag making machinery was purchased unspectacularly as part of the normal routine. Thus in the same year that the patent rights for self opening square paper bags were acquired the firm bought a Patent Cone Bag Machine from Woolley Brothers of Brighton for £110, less 2.5 per cent discount, and a Carter's Patent Canister Tea Bag machine costing £15 was installed in the Hand Bag Making Department.

Overall in the 1890s a further seven bag making machines were installed in the Machine Bag Making Department, of which two were made by the firm's own engineers and three were purchased second hand, making a total investment of over £2,000. As a result the combined output from hand operated and power driven bag machines now exceeded the figure of which the firm had boasted in 1896, a total of 2,700,000 paper bags per week.[23]

Investment in printing equipment was an essential accompaniment to the expansion of paper bag production, for whilst plain bags were still in demand, increasingly retailers saw paper bags not just as a convenient form of packaging their produce but also as an effective means of advertising. By providing printing services, rather than leaving the field clear for local printers in various localities willing to execute such orders, Davidsons hoped to strengthen customer loyalty and increase profit margins, whilst the printing facilities established might in turn generate additional income by attracting orders from other sources. As early as 1886 advertisements in the trade press stressed that the printing of paper bags was done at the mills. Some 10 years later contemporary observers were referring to the extensive printing works at Mugiemoss as a notable feature of the firm's activities and the Printing Department ranked as a major employer of labour with an annual wage bill in 1898 of £2,041 as compared with £2,100 for the Machine Bag Department and £2,346 for the Finishing and Packing Department.[24]

Moreover printing facilities were also provided at other branches. The London warehouse had a separate printing plant and at the Newcastle office an arrangement was made in 1895 with J.S. Bird of Elswick Court Printing Works. He was provided with the latest improved Demy Wharfedale printing machine purchased from the Otley foundry of the Wharfedale Works and he agreed to execute all the printing work the Newcastle office required from time to time, deducting the rent for the machine from his charges each month.[25]

The use of printed paper bags was now expanding rapidly throughout the United Kingdom. The *Paper Trade Review* reported in 1890 that

> Country printers and stationers are cultivating the paper bag trade to a much greater extent than formerly, and even small establishments keep large stocks of assorted bags ready to print to the order of their customers. Improved machinery for the manufacture of paper bags has been introduced on the market during late years, which has given an impetus to the trade owing to economical working.[26]

Meanwhile other paper mills had entered the field so that the *Paper Mills Directory* for 1900 listed 19 paper manufacturers making their own bags as compared with only 5 in 1871. In Scotland the sole competitor in 1871 – Smith, Anderson & Company of Leslie, Fife – had been joined by Oswald & Hall of Bonnybridge near Stirling, Guthrie, Craig, Peter & Company Limited of Brechin, and J.W. Dixon of Markinch, Fife. Nevertheless contemporary observers were of the opinion that C. Davidson & Sons at the end of the

19th century were still the largest makers of paper bags in the United Kingdom.[27]

The growing variety of products available to the firm's customers was further increased by selling papers produced by other paper manufacturers, a practice also adopted by Culter Mills Paper Co. Ltd locally and by some other firms in the industry. The additional revenue helped to cover the overhead costs of the sales organisation and contributed towards the salaries of commercial travellers, whilst there was an additional benefit in that it enabled the firm to offer customers a wider range of goods which in turn might attract additional orders for its own products.[28]

The function of paper merchants developed during this period into a major aspect of the firm's business activities. In the 1890s when the definition of sales in the annual accounts was enlarged to include all sales made by the company the level of income from sales and other sources more than doubled. In 1888–90, immediately before the change, sales averaged nearly £89,000 a year whereas from 1891–93, immediately the new basis was adopted, sales soared to an average of more than £192,000. At the end of the decade when we can, for a short period, measure the importance of sales of other firm's products more directly, sales of Davidsons own manufactures were accounting for approximately £90,000 a year whilst total sales averaged some £173,000. Symbolising the new outlook were the entries in the annual *Paper Mills Directory* for the United Kingdom which after the turn of the century listed C. Davidson & Sons both among the paper manufacturers and separately in the much shorter list of London Wholesale Stationers where it appeared beside such illustrious papermaking names as Alexander Cowan & Sons of Edinburgh, John Dickinson & Co., James Spicer & Sons, and Wiggins Teape.[29]

In the final quarter of the 19th century the firm had thus successfully diversified its operations and now derived income from several different sources whilst retaining paper manufacture at Mugiemoss as the heart of its business. Its rural origins were still reflected in a residual income from farm rents and fishing rights, and some of its paper products had changed little from its early days of machine made paper, but an increasing proportion of its paper was made up into paper bags in all kinds and sizes whilst some of the newer paper products were marketed not in traditional outlets but in the case of felts to furnishing shops and the building trade. The Printing

Department, set up to serve the needs of customers who saw Davidsons paper bags as a means of advertising their goods and their premises, earned income on other orders and the Engineering Department, established to service the firm's own machinery, was also capable of securing an income from external sources. Above all there was the growing income from dealing in all kinds of paper. The varied sources of income offered multiple possibilities for future growth, but even more they offered greater security in an increasingly competitive environment.

CAPITAL EXPANSION

The investment required for diversifying production, increasing output, and enlarging the sales network, was partly self-financed by ploughing back profits into the business. But in addition substantial sums had been raised from outside sources by the end of the 19th century.

The issue of shares to the general public proceeded rather slowly in the years immediately after the firm became a limited company. Only 4,600 shares of £10 fully paid up were issued within the first eighteen months of existence out of the authorised limit of 8,000 shares, including 3,900 shares allotted to the four Davidson brothers under the terms agreed for the sale of the existing business to the new company. By 1882 the issued capital had risen to £60,000 consisting of 5,000 fully paid shares of £10 and 2,000 shares on which £5 had been paid, and the following year permission was obtained to convert all the £10 shares into ten £1 shares, thus making the shares a more marketable commodity and making it easier to raise capital from the general public. Nevertheless it was not until 1888 that the whole of the authorised share capital had been issued, with the issue of the final 10,000 shares of £1 at a premium beginning in the previous year.[30]

Meanwhile money had been borrowed from a few wealthy individuals locally in the form of bonds backed by the security of the firm's property. Bonds on Mugiemoss to the value of £12,000 and on Bucksburn to the value of £4,500 had been granted by July 1877, and three years later a further bond of £1,600 had been taken out on the firm's warehouse and houses in Aberdeen. Later, in 1893, the new London warehouse at 23 Upper Thames Street with the adjacent Paul's Pier Wharf was financed partly by a bond of £3,300 from the Economic Life Assurance Company of London at 4 per cent interest, the loan to be repaid by £200 each year.[31]

In May 1891 the authorised share capital was raised to £100,000 by creating an additional 20,000 shares. These were issued at a 25 per cent premium of 5 shillings per share, the sum obtained being placed in two Reserve Funds, one of which was created specifically for the eventual purchase of the leases of the firm's property at Mugiemoss, Bucksburn and elsewhere. By 31 July 1894 more than £16,000 of these fully paid shares had been issued thus increasing the total issued share capital to £86,366. A further 1,011 shares were issued over the next two years and in 1899 the directors of the Anglo American Self Opening Square Paper Bag Machine and Manufacturing Company, as we have seen, received 1,000 shares as part of their agreement with Davidsons, but the remaining 1,623 shares were never issued.[32]

The Board had in fact resolved in 1896 to take a different approach by issuing debentures. Such issues had become increasingly popular with industrial and commercial companies in the later 19th century since extra capital was raised without handing over any voting rights which could affect the way in which the controlling management group ran the company, whilst the element of gearing introduced in the capital structure meant that the management group holding much of the ordinary share capital benefited, like other ordinary shareholders, more fully from higher profits during prosperous years.[33]

The Davidson directors decided to create 4,000 debentures of £10 bearing interest at four and a half per cent. These were to be issued at a premium of 4 shillings per debenture. The public circular for prospective purchasers in May 1896 stated that the proceeds were to be used to pay off bonds amounting to £20,800 secured on the properties of the company at Mugiemoss, Bucksburn, and elsewhere, with the balance being used to extend the company's business. The issue was an immediate success and by 31 July 1897 the full £40,000 had been raised plus premiums of £800, which paid for the bulk of the costs of the issue, handled by the Omnium Contract Corporation Ltd. of London.[34]

Overall, as can be seen from Table 7, the company increased its capital nearly three fold between 1876 and 1899 from £46,000 to nearly £130,000. £16,000 of the increase represented the balance

Table 7: Issued capital, 1876–1899

Year Ending 31st July £	Paid Up Shares £	Bonds £	Debentures £	TOTAL £
1876	46,000	—	—	46,000
1877	50,000	16,500	—	66,500
1878	50,000	16,500	—	66,500
1879	50,000	16,500	—	66,500
1880	50,000	18,100	—	68,100
1881	50,000	18,100	—	68,100
1882	60,000	18,100	—	78,100
1883	60,000	17,500	—	77,500
1884	60,000	17,500	—	77,500
1885	60,000	17,500	—	77,500
1886	60,000	17,900	—	77,900
1887	66,667	17,900	—	84,567
1888	70,000	17,900	—	87,900
1889	70,000	17,900	—	87,900
1890	70,000	17,900	—	87,900
1891	74,091	17,900	—	91,991
1892	78,183	17,900	—	96,083
1893	82,274	17,900	—	100,174
1894	86,366	21,200	—	107,566
1895	86,366	21,000	—	107,366
1896	87,377	12,000	22,629	122,006
1897	87,377	1,600	40,000	128,977
1898	87,377	1,600	40,000	128,977
1899	88,377	1,600	40,000	129,977

Source: *Annual Report and Accounts*, 1876–1899

of the purchase price of the old firm, leaving a sum of £68,000 for investment in new plant, buildings and additional stocks of materials, or for adding to the firm's reserves to meet future needs.

THE FRUITS OF ENTERPRISE

The environment in which the Davidson directors operated in the final quarter of the 19th century was highly competitive. There were more than 200 paper manufacturing firms in Britain making between them several dozen different products. Few firms attempted to produce the complete range of papers, hence direct competition within the industry was between firms making similar products in the same price and quality range. Nevertheless the degree of competition for most products, measured by the number of manufacturers, was quite strong. In Davidsons' own field in 1900, for example, there were some 80 producers of brown papers and a similar number making various kinds of cartridge papers, some 40 firms making *manilla papers*, and approximately 30 manufacturers of *small hands*, of *middles*, and of *mill wrappers*. Moreover, as was noted earlier, even in the manufacture of paper bags where the firm had been a pioneer, competition had increased considerably so that 18 other paper manufacturers were in direct competition by 1900 as compared with only 4 in 1870.[35]

Fierce competition between the various producers ensured that firms who were determined to stay in business modernised their plant at regular intervals and adopted the latest cost cutting techniques. Costs were reduced by the introduction of esparto grass in the 1860s and wood pulp in the 1880s to supplement or replace the traditional and more expensive raw materials, pulped linen and cotton rags. At the same time chemicals used to treat the raw materials fell appreciably in price, whilst the increasing size of plants – not least wider and faster papermaking machines – produced significant economies of scale.[36]

Continental manufacturers who had access to abundant timber supplies for the production of wood pulp increased the pressure on prices and profits, their products entering the British market unhindered by any import duties. Until the early 1880s paper imports exclusive of millboard, pasteboard, and paper hangings were less than 30,000 tons a year, equivalent to perhaps 10 per cent of the British output of paper. Subsequently imports increased rapidly, stimulated by technical advances which yielded good quality wood pulp made by chemical means to supplement supplies of the poorer quality wood pulp made mechanically and by the widespread adoption of wood pulp as the principal raw material used by Continental paper manufacturers. Imports of paper for printing or writing purposes increased from some 9,000 tons in 1882 to more than 24,000 tons in 1894 whilst imports of other paper 'unenumerated', which included wrapping and packing papers, increased more than five fold from some 15,000 to more than 77,000 tons. In 1895 when the official classification was changed to 'unprinted paper', which embraced both the former categories, the total import exceeded 104,000 tons and by 1899 the total had increased by almost 50 per cent to a new record of 154,000 tons.[37]

The conjunction of falling production costs, a substantial increase in productive capacity and much keener foreign competition led inexorably to lower prices for paper and a squeeze upon profit margins. In the trade journals manufacturers lamented the low level of prices and complained about 'unremunerative trade'. Agreements to limit output and stabilise prices soon ended in failure and the formation of Wall Paper Manufacturers Limited in 1899, created from 31 wallpaper manufacturing firms, was the only significant merger of competing firms.[38]

A.D. Spicer, a member of the well known family of London wholesale stationers, in a perceptive historical survey of the paper industry published in 1907, concluded that the overall price of paper in Great Britain had fallen from an average of 6d per pound in 1860 to 2d per pound in 1902. Using a shorter time scale he thought that the cheapest papers, brown paper and wrapping paper, were generally being sold for less than 1¾d in 1906 as compared with approximately 3½d per pound between 1865 and 1875.[39]

The volume of sales recorded in Davidsons's annual balance sheets reflects the marked fall in paper prices and tends to distort our view of the growth of the firm's business at this period. The figures in Table 8 show a slight decline in gross income which consisted very largely of the revenue from sales of paper, from an annual average of £92,471 in 1876–78 to £88,762 in 1888–90. This was followed by a dramatic increase to an annual average of £192,764 in 1891–93 after the definition of sales was widened to include the re-sale of papers made by other manufacturers. Subsequently sales declined again, averaging £170,903 a year from 1897 to 1899, of which the firm's own papers – as we know from the single surviving Directors' Minute Book – accounted for over £90,000.

It is impossible to determine total paper sales before 1891 but a rough estimate of the volume of sales of Davidsons' own products may be made. If it is assumed that the papers made by the firm declined in price by approximately 50 per cent, which is probably a conservative estimate, then the volume of these sales doubled between 1876 and 1899.

Evidence of the physical expansion of output, though rather sparse, strengthens the belief that the increase in production was

Table 8: Sales and other Sources of Income, *1876–1899*

Financial Year (ending 31 July)	Gross Income[1] £	3 Year Data: Total £	3 Year Data: Annual Average £
1876	94,401		
1877	91,548		
1878	91,465	277,414	92,471
1879	88,599		
1880	93,778		
1881	97,604	279,981	93,327
1882	97,868		
1883	90,165		
1884	93,392	281,425	93,808
1885	92,747		
1886	89,816		
1887	87,185	269,748	89,916
1888	93,178		
1889	88,507		
1890	84,600	266,285	88,762
1891	194,560[2]		
1892	193,025		
1893	190,707	578,292	192,764
1894	187,225		
1895	184,825		
1896	171,103	543,153	181,051
1897	167,019		
1898	165,095		
1899	180,595	512,709	170,903

Source: *Annual Report and Accounts*, 1876–1899

Notes: 1 Sources of income other than paper sales were not indicated separately after 1885. Before this date they were of little importance, e.g. the income from house rents and farming and fishing rights combined never accounted for even one per cent of the total gross income.

From 1886 to 1890 all sources of income were included in the description 'Paper and Sundries' and from 1891 under the phrase 'Total Sales – Paper etc.'.

2 The figures from 1891 onwards were enlarged by the inclusion of the re-sale of paper purchased from other paper manufacturers.

substantial. The paper-making machine with the smallest capacity and the least efficient of the three – the 54 inch machine at Waterton Mill – had been taken out of production at the beginning of the 1880s, but the capacity of the other two machines, capable of making paper 63 inches wide, was increased by various improvements which more than compensated for the closure. Output of paper grew from more than 3,400 tons a year in the early 1870s to more than 6,000 tons a year by the end of the century.[40]

However the signs of growth at Mugiemoss were most tangible in the growing complex of buildings spreading across the site. By 1896 the firm's operations occupied approximately 15 acres as compared with only one acre forty years earlier.[41]

The value of the firm increased correspondingly. In current prices between 1876 and 1899 the fixed assets – after allowing for depreciation each year at the rate of £2,000 – increased in value from £48,252 to £65,047 whilst stocks of raw materials and finished goods rose from £26,754 to £50,740.[42]

However the most convincing yard stick of success is the level of profits generated. It is clear from a study of Table 9 that the firm's operations were consistently profitable.

The best results came between 1876 and 1893. In only one of the first 18 full financial years after the company was formed did profits fall below £7,600. In four of them profits exceeded £10,000 and the average level exceeded £9,200 a year.

The directors adopted a generous dividend policy. In the same 18 years only once did the company fail to pay a dividend of 10 per cent, the level regarded as a reasonable rate of return on capital when the company was formed, and this was balanced by a payment of 15 per cent in 1882.

Moreover a healthy reserve was established to cover future contingencies. £10,100 was set aside between 1876 and 1878 and thereafter for the next eight years £1,000 was added to the Reserve Fund. The Fund was further increased from 1887 to 1891 by £8,092 obtained from the premium on the issue of new shares.

On various stock exchanges the company's shares traded at a substantial premium. In the summer of 1887, for example, a Manchester Stockbroker who dealt in the shares of several paper manufacturing firms was quoting the fully paid £1 Davidson shares at a price of £2 and the £1 shares on which 10 shillings had been

Table 9: Profits, Dividends and Reserve Funds, 1876–1899

Year Ending 31 July	Net Profits[1] £	Dividends £	Dividends Per Cent	Sums added to Reserves: Reserve Fund £	Sums added to Reserves: Leases redemption Fund[2] £
1876	10,442	3,153	7.5	4,300	—
1877	9,753	4,800	10.0	3,800	—
1878	8,564	5,000	10.0	2,000	—
1879	8,563	5,000	10.0	1,000	—
1880	7,642	5,000	10.0	1,000	—
1881	7,902	5,000	10.0	1,000	—
1882	12,824	9,000	15.0	1,000	—
1883	9,090	6,000	10.0	1,000	—
1884	9,408	6,000	10.0	1,000	—
1885	9,241	6,000	10.0	1,000	—
1886	9,707	6,000	10.0	1,000	—
1887	9,408	6,000	10.0	3,333[3]	—
1888	8,670	7,000	10.0	1,667[3]	—
1889	9,240	7,000	10.0	—	—
1890	6,451	7,000	10.0	—	—
1891	10,889	7,000	10.0	3,092[3]	1,000[3]
1892	10,054	7,818	10.0	—	200
1893	8,011	8,230	10.0	—	200
1894	6,719	4,319	5.0	111	200
1895	7,813	5,182	6.0	79	200
1896	9,044	6,550	7.5	262	200
1897	7,103	6,553	7.5	—	200
1898	6,258	4,369	5.0	—	200
1899	6,553	4,402	5.0	—	200

Source: *Annual Report and Accounts*, 1876–1899, Private Journal No. 2, 1895–99

Notes: 1 Surplus available, after meeting all expenses, for depreciation, payment of dividends, placing in reserves, and carrying forward to next year. Hence directors' fees and interest on loans, bonds secured on property and debentures have been deducted.

2 Excludes the annual interest earned on debentures and other securities purchased by the Fund.

3 Paid out of the premium on the issue of new shares.

paid up at £1. The favourable opinion of both actual and potential investors in the firm was also reflected in the relative ease with which extra capital was raised at a premium in 1891 and 1896.[43]

During the later 1890s the company's results were less

impressive. After 1893, in four of the next six years profits fell below £7,600 and the annual average of £7,248 was 21 per cent lower than in the previous 18 years. Dividend payments reflected this decline, ranging from 5 per cent to 7 and a half per cent, and less than £2,000 was set aside for reserves out of profits.

The poorer performance provoked some expressions of discontent from shareholders who were concerned not only about lower dividends but also the decline in share prices, which would mean a capital loss should they wish to realise their holdings. Letters of complaint from two shareholders were precipitated by a proposal in February 1898 to alter the Articles of Association of the company. John Davidson, one of the original directors named in the Articles, had died in November 1897 thus reducing the number of directors to only two, and it was proposed that John Mackie, a manager of the London office, Charles William Davidson, the son of Charles Davidson Junior who died in 1870, and William Dalzell Davidson, the son of Alexander Davidson, should be admitted as directors, and their names listed, together with the two surviving directors, in a new clause 46 of the Articles which was to replace the original one. A further section of the new clause – 46(d) – stipulated that a person wishing to stand as a director, unless recommended by the existing directors, had to be nominated in writing to the company secretary at least seven clear days before the meeting at which he was standing for election.[44]

George Bruce of Elgin declared that, 'I consider it my duty to lodge my decided protest against the appointment of either Mr C.W. Davidson or Mr W.D. Davidson as Directors. We have too many Davidsons already, under their management we have lost nearly one half our capital invested in the concern were our shares to be sold at present prices. This is no light matter to the shareholders, consequently my confidence in their management is gone. I request that you will read this letter at an early stage of the meeting to be held on 24th inst.'

Robert S. Cumming of Aberdeen, who strongly objected to the proposed new clause 46(d) observed that, 'Some of the Directors proposed are quite young men, and their experience cannot amount to much. Moreover, I understand some of them reside in London and it must be a very expensive thing their coming down to attend meetings in Aberdeen. Would it not be much more to the point to appoint some of the larger shareholders residing in the district, men of business experience and who could attend the meetings without expense to the Company and without much loss of time to themselves.'

He returned to the attack in September, complaining about the dividend of only 5 per cent.

> Compared with Culter Mills Company Limited and former years this looks bad; and may account for the reason of want of confidence as regards the share market by the general public, *no rise*. It has been hinted that the concern will never be a good thing for the general shareholders so long as the present Directors and Managers are in power, as they require too big a slice to themselves in the shape of large salaries and other expenses. Now if there is any truth in this, it will require to be looked into at Annual General Meeting, and compared with other works of similar kind, and if necessary rectified.

The alteration of the Articles of Association was duly approved, the three directors proposed were elected to the Board, and next year's dividend was again at 5 per cent. Whether the Board's future behaviour was influenced by the protests in other ways we shall never know.[45]

However although the shareholders' concern is understandable, if we take a broader perspective even the weaker results of the late 1890s seem quite respectable. Investors in British industrial and commercial firms generally experienced a secular decline in the real rates of return on their capital during the period of consistently falling prices from 1873 to 1896 which has been labelled 'the Great Depression'.[46] The profits and dividends of Culter Mills Paper Company Limited, regarded with such favour by Robert S. Cumming, were also declining, albeit from a higher level, and within the Aberdeen area not far from Mugiemoss was a more striking example of the pressures acting upon the British Paper Industry. Alexander Pirie and Sons of Stoneywood, after a sparkling performance in the 1870s and most of the 1880s, could only manage to pay a dividend as high as 7½ per cent on two occasions during the 1890s and even after a very substantial investment programme and re-structuring of the company in 1898 the dividend paid on the ordinary shares never exceeded 6 per cent.[47]

Chapter 3

Crisis and Recovery, 1900–1939

MOUNTING PROBLEMS, 1900–1914

The more modest level of profits and dividends continued in the early years of this century. After payment of debenture interest of some £1,700 a year and all other expenses, profits recorded in the directors' annual report to shareholders averaged £5,521 a year in the seven years 1900–1906, reaching a peak of nearly £8,000 in the boom year of 1900 and declining to just under £4,000 in the depression of 1903. Dividends were paid at the rate of 5 per cent from 1900 to 1902 and then ranged between 3.0 and 4.5 per cent.

After 1906 the situation deteriorated appreciably. Profits after paying interest on the debentures fell to less than £900 in 1907, recovered to nearly £1,600 the next year, and in 1909 there was an actual loss of £785. Not surprisingly no dividends were paid over the three years.

The board of directors reacted to the worsening situation in several ways. As early as 1903 their printed report to shareholders issued before the Annual General Meeting had begun to include more explanation of the year's results and the policies being pursued. In 1908 in a further move to inspire confidence and also to gain the benefit of outside advice and connections, Alexander Marr, a director of the neighbouring firm of Alexander Pirie & Sons Ltd., which was probably the largest papermaking firm in Scotland, was persuaded to join the Board. He had been secretary of that firm for 26 years and had nearly 40 years' experience in the industry. His appointment to the Davidson board required an alteration to the Articles of Association, raising the maximum number of directors from five to six, and was a major break with tradition since he was the first director to be appointed

Table 10: Profits and Dividends, 1900–1914

Year Ending 31 July	Net Profits[1] £	Dividends £	Per Cent
1900	7,946	4,419	5.00
1901	5,164	4,419	5.00
1902	5,023	4,419	5.00
1903	3,999	3,535	4.00
1904	5,137[2]	2,651	3.00
1905	6,109	3,977	4.50
1906	5,267	3,535	4.00
1907	899	—	—
1908	1,579	—	—
1909	−785	—	—
1910	3,186	2,081	2.50
1911	1,822	1,040	1.25
1912	892	—	—
1913	1,959	1,040	1.25
1914	101	—	—

Source: *Annual Report and Accounts*, 1900–1914; Private Journal No. 2, 1900–1914.

Notes: 1 Surplus available, after meeting all expenses, for depreciation, payment of dividends, placing in reserves, and carrying forward to next year. Hence directors' fees and interest on debentures, loans and bonds secured on property have been deducted.

2 Includes £ 941 profit received from the sale of property in London.

who was neither a member of the Davidson family nor a previous employee.[1]

Further re-assurance was provided by the conversion of the leases over the land and buildings at Mugiemoss and Bucksburn into a permanent feu. The idea had originated several years earlier when a Leases Redemption Fund had been started with £1,000 allocated from the premium on the issue of new shares in 1891. Each year a further sum was added, usually in excess of £200, and by 1906 the Fund had accumulated more than £5,000. However the realisation that substantial new investment was needed at Mugiemoss and Bucksburn sharpened interest in converting the leases. Negotiations with the landlords, Alexander Pirie & Sons Ltd. commenced in 1907 and the feu charter was finally granted in

1911, the balance in the account after meeting the costs of conversion being transferred to the newly created Special Reserve Fund, to be used for other purposes.[2]

Meanwhile steps were taken to improve the sales network. More than £4,000 was spent between 1907 and 1909 on acquiring and improving a freehold warehouse in North Frederick Street, Glasgow, to replace a smaller leasehold property in Howard Street, which had been used since vacating the premises in St Enoch Square. A new improved location was also obtained for the firm's Edinburgh operations which now centred on a rented warehouse in Borough Loch Square in place of one in Castle Terrace. In London some £3,000 was invested at Paul's Pier Wharf, 23 Upper Thames Street, the principal object of expenditure being an extension of the Printing Works.[3]

However the major new initiative taken to revive the firm's fortunes was the decision to diversify into the manufacture of better quality papers with a higher profit margin. The two papermaking machines at Mugiemoss could not be adapted readily to make the new products hence it was decided to invest an estimated £12,000 to £15,000 on new machinery for this purpose.

The new strategy was set out in the directors' annual report and in a special circular to shareholders sent out shortly before the Annual General Meeting in September 1907 when, for the first time since the company's formation in 1875, no dividend was declared. The firm's difficulties were attributed mainly to severe competition in the production of the more traditional heavy wrapping papers and in particular to a rising tide of cheap imports made from wood pulp as a result of which profit margins had been squeezed severely in the previous 2–3 years.

The directors might have added that output had been falling since the turn of the century. As can be seen from Table 11 output of paper exceeded 6,700 tons each year from 1898–99 to 1900–01, declined by rather more than 10 per cent over the next four years, and then dropped sharply to barely 4,700 tons in 1908–09.[4]

The heart of the programme was a third papermaking machine ordered from Bertrams Ltd. of Edinburgh, with engines provided by Ashworth & Parker of Bury, in 1908. This was a large single cylinder (or Yankee) machine making machine glazed paper 108 inches wide as compared with paper 63 inches wide produced on the firm's other two papermaking machines. The wet end of the

Table 11: Output of Paper, 1898–1914

	Paper Output	
Year Ending 31 July	Tons	Index Number (1914 = 100)
1899	6,804	110
1900	6,701	109
1901	6,708	109
1902	5,748	93
1903	5,976	97
1904	5,829	95
1905	5,807	94
1906	6,073	99
1907	4,973	81
1908	5,407	87
1909	4,702	76
1910	5,912	96
1911	5,480	89
1912	5,863	95
1913	6,396	104
1914	6,194	100

Source: Consumption of Coals, Mugiemoss, monthly summaries 1895–1923 (with annual paper production).

machine on which paper was formed from pulp, was similar to that of the normal Fourdrinier machines used at Mugiemoss and throughout the United Kingdom, but at the dry end of the machine the series of drying cylinders was replaced by a single large drying cylinder. The outside of the cylinder was highly polished and the paper passing round this cylinder was dried and glazed on one side.[5]

The MG machine, including delivery and erection, cost more than £4,200 and a total of nearly £9,000 was spent on additional plant at Mugiemoss and Bucksburn in the financial year ending 31st July 1909 during which the machine began operations. However associated changes in steam generating equipment partly or wholly connected with the new machine increased the overall sum invested. A new 'first class' steam engine was installed and running by August 1911 at a cost of nearly £3,000 and in the following two years the steam boiler plant was extended and improved, entailing expenditure of nearly £1,200 on the new boiler alone. The installation of two new pulping engines in the same two years thus raising the raw material input, completed the project.[6]

The effect of this investment was a substantial increase in the maximum productive capacity of the mills, but another objective was a reduction in costs through increased efficiency. In particular the improved boiler plant and steam engine, aided by the earlier introduction of a super heating mechanism in 1904–05, meant economies in the use of coal, which was increasing in price.[7]

The most tangible fruit of the investment nevertheless was seen in the quality and variety of products. The firm already made some glazed papers 63 inches wide, in a two stage process, the glazed surface being produced after the paper had left one of the existing paper making machines by passing it through a supercalender or a friction calender. The new machine, making glazed paper 108 inches wide directly on the machine itself, substantially increased the output of glazed paper. The relatively thin lightweight texture of the paper, combined with toughness and flexibility found a ready use in improving wrapping papers and paper bags in the firm's product range whilst other products were introduced yielding potentially higher profit margins and designed to attract new customers or recapture former ones. The firm's policy of using its large sales network to re-sell papers bought from other paper mills here proved its worth in providing some 2 or 3 years' experience of marketing such products before making them in its own mills.[8]

Among the new products were papers for packing starch, tea and pipe tobaccos, and cheap envelope papers – the glazed side of the paper being suited to writing upon and the rough side taking the gum for the flap. Poster papers, lithographic papers used to print illustrations, *sealings* – paper for wrapping parcels fastened with sealing wax, which adhered readily to the rough side of the paper – and cheap buff paper used by railway companies for forms and envelopes, hence the term *railway buffs*, gave further variety.[9]

The search for lightness combined with strength resulted in a growing production of Kraft Paper made from chemically treated wood pulp and marked a major shift in the composition of the fibrous raw materials used for paper-making at Mugiemoss. The process originated in Sweden in the early years of this century and the extremely tough thin brown paper, equivalent in strength to several sheets of thick paper of ordinary quality, found a ready market abroad, stimulating its production in other countries. The first two British producers recorded in the Paper Mills Directories were listed in 1905. Davidsons began production on its new machine and by 1913 it was one of 18 firms in the United Kingdom engaged in Kraft production.[10]

The various measures taken by Davidson's directors to reverse the downward trend in profits were not completed until 1913. A year later the outbreak of the First World War began to change the economic parameters upon which the programme was based. Hence any judgement upon the effectiveness of the programme must be very tentative.

The volume of output recovered from the low point reached in 1908–09, when barely 4,700 tons of paper were produced, and on the eve of the First World War output averaged nearly 6,300 tons a year. However, this was still below the level achieved from 1898 to 1901 when only two paper making machines were in operation and there is no doubt that the Paper Making Department of the mill, after the installation of the third machine in 1909, was working at substantially less than its full potential.

The overall level of gross income, which was derived mainly from sales of paper products, as shown in Table 12 was slightly lower in the years before 1914, despite the move up-market to increase the volume of products which would command a higher price and profit margin. The directors' annual reports indeed commented on the unsatisfactory level of paper prices on several occasions between 1910 and 1914 which reflected the severe competition from other British and foreign producers. At the same time rising costs, particularly coal and freight charges, meant a squeeze on profit margins, whilst the introduction of National Health and Unemployment Insurance, though requiring a modest £240 a year, added to the pressure.[11]

Profits recorded in the directors' annual reports recovered from the nadir of 1907–09, reaching a healthy £3,186 in 1910, when a dividend of 2.5 per cent was declared, ending the three year famine. But in the next four years profits never exceeded £2,000 and in two of them the level was below £900. Dividends were passed on two occasions and in the other two years the dividend was a mere 1.25 per cent.

Table 12: Sales and Other Sources of Income, 1900–1914

Year Ending 31 July £	Gross Income[1] £
1900	188,726
1901	187,961
1902	177,897
1903	180,967
1904	182,564
1905	187,118
1906	186,925
1907	181,442
1908	183,328
1909	171,017
1910	174,908
1911	180,014
1912	176,411
1913	180,884
1914	178,727

Source: *Annual Report and Accounts*, 1900–1914.
Note: 1 Defined as 'Sales – Paper etc.' for the years 1900–1904, from 1905 onwards as 'Sales, Dividends, Interests in Other Companies'. It is clear from Private Journal No. 2, 1895–1914, that income from interest and dividends, house rents, and farming and fishing rents accounted for barely 2 per cent of the total each year.

WARTIME PROFITS AND THEIR DISPOSAL, 1914–1922

The outbreak of the Great War in 1914 greatly changed conditions for the British Paper Industry. The most immediate effect was a shortage of manpower.

Most people, including the employers, the Liberal Government and the Conservative Opposition, assumed that the war would be decided quickly by a few decisive battles. No obstacles were placed in the way of men volunteering to join the Armed Forces and in a surge of patriotic fervour both skilled and unskilled men rushed off to enlist.[12]

At Mugiemoss there was practical inspiration at the highest level in the example of Thomas Davidson, the youngest director, who had been a member of the Territorial Army since his days as a student. He left Mugiemoss at the outbreak of the war, took charge of the First Battery of the First Highland Brigade soon afterwards, and served for a year in France, winning the D.S.O. and the Croix de Guerre with palm for gallantry.[13] By October 1914 so many men from the factory had enlisted, including skilled operatives, that a third of the papermaking machinery stood idle, a situation which continued into the summer of 1915 and beyond. Throughout the war in fact the directors continued to comment that machinery was at a standstill and the mill was working short time because of the absence of so many men in the Army. Overall a total of 179 men had joined the Armed Forces by the summer of 1918, more than 100 of them from Mugiemoss itself, out of some 700 persons employed by the firm in the years immediately before the outbreak of war.[14]

Production at Mugiemoss and other paper mills was also affected by a growing shortage of raw materials. Imports of rags, esparto and wood pulp continued for several months virtually without official restrictions but by 1916 the increased demands on the available shipping capacity, not least the transport of troops and munitions, led to direct government control. In February 1916 the Royal Commission on Paper was established to impose some control over the manufacture, sale and use of paper. Licences were required from 1st March 1916 for the import of paper and paper-making materials. The first year of control reduced the combined import by 36 per cent compared with the figure for 1914 whilst in the following year ending 28th February 1918 more severe restrictions cut imports of paper-making materials to 38 per cent and imports of various kinds of paper to less than 22 per cent of the 1914 tonnage. Home supplies of waste paper, the collection of which was organised by many local authorities encouraged by government circulars, and of straw, which was more plentiful as cereal growing expanded to reduce the need for grain imports, were insufficient to fill the gap and overall paper production fell below the 1914 level.[15]

Demand for paper was buoyant. In addition to the normal peacetime requirements, swollen by rising real incomes, fuller employment, and a decline in imported supplies of paper, paper was used in a variety of products for the Armed Forces ranging from torpedo tubes, fuse boxes, cartridge cylinder linings and smoke boxes to cardboard containers for packing biscuits, jam and other food for the Army. Despite attempts by the Government to restrict non essential uses of paper, severe shortages developed, resulting in a steep rise in prices which more than kept pace with the rising prices of raw materials, thus boosting manufacturers' profits.[16]

The directors of Davidsons, struggling to cope with shortages of labour and materials, were able to maintain production at a level some 30 per cent lower than in the last year of peace. In 1916–17 output of paper fell to less than 4,500 tons as compared with nearly 6,200 tons in 1913–14, a decline of 27 per cent, whilst in the best year 1915–16 output was only 14 per cent below the pre-war level.

In the circumstances the directors must have found considerable consolation in the firm's financial results. Income from sales of paper products and from other sources increased from nearly £179,000 in 1913–14 to more than £460,000 in 1917–18 despite a lower level of production. Profit margins widened and profits recorded in the directors' annual reports to shareholders increased from £101 in 1913–14 to nearly £26,000 in 1917–18, a record figure for the firm. Overall in the four years of war annual profits averaged £17,254 – a figure which also broke all previous records. The level of dividends paid to shareholders reflected the new found prosperity, rising to 10 per cent in 1915–16 then to 12.5 per cent and 15 per cent in the final year of the war.[17]

Satisfaction with these results must have been qualified by the reflection that these were figures swollen by wartime inflation. Profits were also boosted by the virtual cessation of investment in new buildings and machinery. This investment had averaged £2,000 a year in the four years preceeding the outbreak of war whereas the total expenditure during the war years barely exceeded £1,000. However the opportunity was taken to pay off the £2,200 mortgage on the Glasgow Warehouse in two instalments in 1915 and 1918.

Of more importance in distorting the overall picture of the firm's performance in the war years and the immediate post war ones was the substantial increase in taxation. The general level of direct taxation was raised, the 'normal' rate of income tax increasing from 14d in the £1 in 1913–14 to 6 shillings in 1918–19 and 1919–20. In addition an Excess Profits Duty payable on the excess of profits above a pre-war standard figure was introduced in 1915 and lasted until 1921. C. Davidson & Sons Ltd. had paid out more than £34,000 in Excess Profits Duty between July 1917 and July 1919, nearly £20,000 of this in the first post war year, offset by a refund of

Table 13: Income, Profits and Dividends, 1915–1922

Year Ending 31 July	Gross Income[1] £	Net Profits or Losses[3] £	Dividends £	Dividends Per Cent
1915	178,518	1,195	—	—
1916	248,865	16,527	8,837	10.0
1917	290,361	25,396	11,047	12.5
1918	460,739	25,899	13,257	15.0
1919	390,952	16,977	13,257	15.0
1920	467,171	15,610	13,257	15.0
1921	396,690[2]	−6,668	2,209	2.5
1922	266,236	−236[4]	—	—

Source: *Annual Report and Accounts*, 1915–1922

Notes: 1 Defined as 'Sales, dividends, interest etc.'

2 Included an estimated sum for the refund of Excess Profits Duty.

3 Surplus available, after meeting all expenses, taxation, and investment in new plant, for meeting depreciation, payment of dividends, placing in reserves and carrying forward to next year's Profit and Loss Account.

4 In addition £14,539 (taken from the Special Reserve Fund) was written off the value of stocks of paper and raw materials.

over £6,000 in 1921–22. Other taxes amounted to more than £16,000 from 1920 to 1922, more than half of this in 1922, whereas in the immediate pre-war years the total tax burden was less than £400 a year.[18]

For nearly two years after the end of the war in November 1918 the price of paper products remained high and profits were buoyant. It was several months before shortages of manpower and materials were overcome and government restrictions on the import of paper were not removed completely until September 1919. However paper mills generally had reached their pre-war levels of production by Autumn of 1919 and as the boom in the economy which had been gathering force since the spring approached its peak demand for paper rose sharply. Mill Managers struggled to cope with the flood of orders and the prices of paper products increased beyond even the high level reached by the end of the war.[19]

The directors and shareholders of C. Davidson & Sons shared in the general prosperity. Gross income from all sources fell below £400,000 in 1918–19 but reached a new record of more than

Table 14: The Burden of War Taxation

	Taxes Levied					
Year Ending 31 July	Excess Profits Duty £	Corporation Profits Duty £	Other Taxes £	Total £	Tax Refunds[1] £	Net Tax Paid £
1911	—	—	311	311	—	311
1912	—	—	385	385	—	385
1913	—	—	374	374	—	374
1914	—	—	379	379	—	379
1915	—	—	425	425	—	425
1916	—	—	543	543	—	543
1917	1,992	—	258	2,250	2,163	87
1918	12,361	—	327	12,688	3,285	9,403
1919	19,776	—	373	20,149	2,209	17,940
1920	356	—	3,404	3,760	—	3,760
1921	—	873	2,137	3,010	—	3,010
1922	112	9	9,920	10,041	6,167	3,874

Source: Private Journal No. 2, 1911–1914; Private Journal No. 3, 1915–1922
Notes: 1 The refund in 1922 was of Excess Profits Duty, the other refunds were of income tax.

£467,000 in the following year. Profits recorded in the directors' annual reports in both years exceeded £15,600 and the dividend level of 15 per cent in 1917–18 was maintained.

Profits would have been still higher but for the substantial expenditure on new plant and repairs in an endeavour to make good the enforced neglect of the war years. Repairs and renewals, which had averaged some £4,000 a year between 1914–15 and 1917–18, rose to £7,900 and then to an unprecedented level of nearly £16,000 in 1919–20. In the same two post war years investment in new plant amounted to more than £11,000. Part of this was spent on conversion to electric power, the current being supplied by Aberdeen Corporation's power station.[20]

Any euphoria which shareholders and directors felt at the Annual General Meeting in September 1919 and September 1920 must have been dispelled by the dramatic change in the general economic climate. By the middle of 1921 Britain was experiencing one of the worst depressions in its history with 22 per cent of all workers insured under the national unemployment benefit scheme recorded as unemployed. The paper industry was especially depressed. The position was so serious that the Interim Industrial Reconstruction Committee held an enquiry into the state of trade and the lack of employment in the industry with a view to making representations to the government. Several months later the *Papermaker* reviewing the situation concluded that:

> 1921 will probably rank as one of the most unfortunate periods ever recorded in the history of the industry . . . so depressed were conditions in the British Isles that many mills were closed or working short time at Christmas and the New Year onwards, and it was well into June (1921) before unemployment showed a slight improvement.[21]

The financial problems facing paper manufacturers were compounded by a sharp fall in prices. Stocks of paper and raw materials had to be written down in value and contracts for delivery of paper-making materials, often placed as much as a year ahead, had to be re-negotiated on the best terms that could be obtained.[22]

The impact at Mugiemoss was severe. A loss of £6,668 was made in the financial year ending 31 July 1921 despite the provision made in July 1920 to carry forward £11,492 of profits in the Profit and Loss Account from the previous financial year. An interim dividend was paid in April 1921 equivalent to 2.5 per cent for the whole year, but the directors wisely decided not to declare a final dividend, normally paid in October.[23]

There was a loss of £236 in the next financial year, and a further £14,539 – taken from the Special Reserve Fund – was written off the value of stocks of paper and raw materials to take account of the drastic fall in prices. In the circumstances it is not surprising that the shareholders received no dividend.[24]

THE STRUGGLE TO SURVIVE, 1923–1934

The general revival of the economy over the following years from 1922 to 1925 was reflected in the steady recovery of the paper industry. Reviewing conditions and developments in the industry during 1925 the *Papermaker and British Paper Trade Journal* was of the opinion that, with one significant exception, 'the close of the year finds the British paper industry in a much better position that it has been since 1913'.[25]

The exception was the section of the industry producing wrapping and packing papers, products which accounted for a major part of the Davidson business. Wrapping and packing papers made in Germany and Scandinavia were often appreciably cheaper than the comparable British product and imports were rising rapidly. In 1924 imports exceeded by some 15,000 tons the pre-war record figure of 201,000 tons reached in 1913 and the upward trend continued into 1925. British output on the other hand declined from an estimated 260,000 tons in 1913 and 300,000 tons in 1920 to some 180,000 tons in 1924. Several paper mills had closed down and approximately a quarter of the papermaking machines in this section of the industry in 1920 were no longer operational by the end of 1924. In addition short time working was widespread and statistics collected by the trade association of wrapping paper manufacturers indicated that on average papermaking machines were standing idle for fifteen and a half weeks a year during 1923 and 1924.[26]

At Mugiemoss short time working was still more marked. A volume which has survived recording the daily output of each papermaking machine from September 1923 to June 1925 reveals the extent of under utilisation. In 1924, as can be seen from Table 15, No.1 machine was operating for approximately 60 per cent of the time, exclusive of Sundays, machine No.2 for only 46 per cent, and No.3 machine a mere 35 per cent.

Short-time working was just as prevalent in the Autumn of 1923. No.1 machine was at work for only 54 per cent of the possible number of shifts between 6 September and 31 December, exclusive of Sundays, No.3 machine for 44 per cent and No.2 machine for 23 per cent. In the first half of 1925 there was a slight improvement with No.1 machine operating for 64 per cent of the available time and the other two machines for 46 and 26 per cent respectively.

The output of paper from the three machines in 1924 amounted to less than 4,500 tons. The maximum productive capacity of the machines at this date is not known but it is clear that even before the M.G. papermaking machine was in operation in 1909 the other two machines were capable of producing more than 6,000 tons a year.[27]

The financial consequences of fewer orders and expensive

Table 15: Short Time Working in 1924

Period Covered	No. of Shifts Papermaking Machines were idle			Maximum No. of Shifts Possible Per Machine[1]	Proportion of Shifts idle Per Cent		
	No.1	No.2	No.3		No.1	No.2	No.3
7 Jan–2 Feb	45	45	45	72	63	63	63
4 Feb–1 March	35	36	48	72	48	50	67
3 March–29 March	31	45	33	72	43	63	46
31 March–26 April	27	54	45	72	38	75	63
28 April–24 May	46	60	35	72	64	83	48
26 May–21 June	12	28	55	72	17	39	76
23 June–19 July	11	24	8	72	15	33	11
21 July–16 Aug	33	57	47	72	46	79	65
18 Aug–13 Sept	48	36	36	72	67	50	50
15 Sept–11 Oct	25	60	48	72	35	83	67
13 Oct–8 Nov	20	44	46	72	28	61	64
10 Nov–6 Dec	19	50	31	72	26	69	43
8 Dec–31 Dec	25	63	39	72	35	88	54
TOTAL	377	602	516	936	40	64	55

Source: Daily Report Books, Volume I, 1923–1925
Notes: 1 Based on the assumption of 3 shifts per day for a six day week, excluding holidays.

short-time working, which may have increased production costs for the typical wrapping paper manufacturer by £2 a ton, were very apparent in the annual accounts received by Davidsons shareholders.[28] As can be seen from Table 16 a substantial loss was made in each year from 1923 to 1927, in three of them exceeding £10,000, and the combined deficit for the five years amounted to more than £57,000. £7,500 was lost in 1924 as a result of the financial problems of the Bay Sulphite Company of Canada in which a number of British papermaking concerns had invested with a view to securing their long term supplies of wood pulp. The bulk of the deficit however was incurred in normal trading operations.[29]

The firm's reserve funds, amounting in 1922 to £24,000 in the main Reserve Fund and £10,000 in the Special Reserve were exhausted three years later, whilst bank loans which stood at £31,195 in 1922 rose to nearly £62,000 in 1926. Continuance of credit was made conditional upon the North of Scotland Bank obtaining further security and in September 1925 the Davidson

Table 16: Income, Profits and Dividends, 1923–1934[1]

Year Ending 31 July	Gross Income[2] £	Trading Profit[3] £	Interest & Dividends Received £	Total £	Interest Payable on loans & debentures £	Net Profit or Loss[4] £	Dividend £
1923	245,589	—	—	—	1,800	-5,666[5]	—
1924	238,338	—	—	—	1,927	-17,499[6]	—
1925	240,681	—	—	—	2,188	-13,714[7]	—
1926	232,290	—	—	—	2,781	-11,455	—
1927	214,828	—	—	—	2,380	-9,124[8]	—
1928	—	5,081	207	5,288	4,820	-10,710[9]	—
1929	—	3,351	204	3,555	4,982	-1,427	—
1930	—	4,289	210	4,499	5,057	-657	—
1931	—	-10,090	200	-9,890	4,435	-14,425	—
1932	—	9,306	203	9,509	5,044	4,496[10]	—
1933	—	3,553	205	3,758	2,356	1,575[11]	—
1934	—	2,932	158	3,090	2,349	-3,140[12]	—

Source: *Annual Report and Accounts*, 1923–1934, Private Journal No. 3, 1923–1932.

Notes: 1 In June 1931 a wholely owned subsidiary company, Davidsons Paper Sales Limited, was formed to run the sales organisation. The figures recorded after that date are therefore not strictly comparable with the earlier figures in the Table.

2 Defined as 'Sales, dividends, interest etc'. No figure was given in the accounts after 1927.

3 This item appeared in the accounts for the first time in 1928.

4 A loss was made in every year except 1932.

5 £5,000 was taken from the Special Reserve Fund to cover the bulk of the deficit, part of which was incurred because of the depreciation in the value of stocks.

6 Included £7,500 written off shares in the Bay Sulphite Company of Canada when the firm collapsed. £12,000 was taken from the Reserve Fund and £5,000 from the Special Reserve Fund to cover the bulk of the loss.

7 £12,000 was taken from the Reserve Fund to cover the bulk of the deficit.

8 £3,000 was recovered in Excess Profits Duty so that the cumulative deficit on the Profit and Loss Account increased by £6,124.

9 The modest surplus of £468 on current operations was turned into a large deficit because special provision of £11,178 was made to provide for the loss on the realisation of the Australian business (the Colonial Paper Company Limited) and to write down the value of stocks in hand and of certain investments of the company.

10 Includes £31 profit by the subsidiary company.

11 The surplus after deducting interest payments was turned into an overall deficit because of a loss of £2,877 by the subsidiary company.

12 The surplus after deducting interest payments was turned into an overall deficit because the subsidiary company made a loss of £3,393. In addition £388 was deducted to pay part of the costs of installing an improved system of accountancy and production control.

directors assigned to the Bank 20,000 partly paid up £1 shares issued in 1881 on which 10 shillings per share was uncalled.[30]

In the midst of the crisis in the Spring of 1926 the most experienced director and the oldest, Alexander Marr, collapsed under the strain. He temporarily lost the use of his legs and after his doctor advised him 'to give up all kinds of business and rest' he tendered his resignation from the Board. Less than six months later W.D. Davidson, based at Mugiemoss, and A.J. Davidson of the London office, in the briefest of letters also resigned as directors.[31]

The other two directors, Charles William Davidson in London, and at Mugiemoss Colonel Thomas Davidson, who had seen distinguished service during the war, decided to weather the storm. They were joined by a complete outsider, Alexander Thomson Dawson, who was appointed to the Board in the same month in which the other two Davidson directors had resigned. He appears to have had no previous experience in the paper industry and his career had been spent as an employee of Morrison's Economic Stores, an Aberdeen firm retailing cheap clothing – popularly known as 'Raggy Morrisons' – of which he became one of the three directors in 1924. Colonel 'Tom' Davidson remained the dominant figure at Mugiemoss but Dawson's new ideas and fresh capital were additional assets in the fight for survival.[32]

The prospects of overcoming the firm's financial problems meanwhile had received a vital boost from government action over import duties. The wrapping and packing manufacturers section of the Papermakers' Association had applied to the Board of Trade in 1925 under amended, more favourable, regulations governing the Safeguarding of Industries Act for a duty to be imposed on these types of paper. The committee appointed to examine the situation after a detailed examination reported in favour and on 1st May 1926 an *ad valorem* duty of sixteen and two thirds per cent was imposed to last for five years. When the duty lapsed in May 1931 imports of wrapping and packing papers again entered Britain unhindered but the landslide victory of a national government at the General Election in October led to the adoption of a general policy of protection before the end of the year.[33] Henceforth imports of wrapping and packing papers were protected by import duties at varying levels from a minimum of 15 per cent to a maximum of 50 per cent *ad valorem* until the outbreak of the Second World War.[34]

The duty imposed in 1926 was less than wrapping paper manufacturers had demanded but the degree of protection afforded was sufficient to promote a revival of that section of the industry, assisted also by growing demand for packaging as the economy generally expanded from 1927 to 1929. Production and employment increased and there was substantial investment in new plant of which the most spectacular example was the construction at Aylesford in Kent of one of the largest factories in Europe to manufacture kraft wrapping paper. When the results of the Census of Production in 1930 were published the extent of the recovery since the previous Census of Production in 1924 was quite striking. Production of wrapping and packing paper, exclusive of oiled, waxed and waterproof wrappings, had increased from 159,000 to 202,000 tons, whilst imports retained in the United Kingdom had declined from 206,000 to 171,000 tons. Meanwhile the share of the home market held by British producers had risen from 38 to 53 per cent.[35]

At Mugiemoss it is not surprising that there were no immediate plans for large scale investment in new plant. It is apparent from Table 17 that less than £2,000 a year, exclusive of repairs and renewals, was spent on new buildings and equipment over the four financial years ending on 31 July 1930.

The most important project was the conversion of No.2 Fourdrinier Papermaking Machine into a single cylinder machine for making machine glazed paper. The decision was precipitated by the passing of the Sale of Food (Weights and Measures) Act, 1926,

Table 17: Investment in New Plant, 1923–1934

Year Ending 31 July	Value of Additional Plant £
1923	3,206
1924	2,387
1925	1,761
1926	4
1927	1,785
1928	1,476
1929	2,594
1930	1,452
1931	713
1932	4,325
1933	4,833
1934	4,272[1]

Source: *Annual Report and Accounts*, 1923–1934

Note: 1 £2,388 of this was specifically for 'Expenditure on Mechanisation of Administration and Improved Production Control'.

which stipulated a weight limit for the paper in which the goods such as bacon, butter, lard and sugar were wrapped and thereby stimulated still further the growing demand for thin lightweight paper possessing toughness and flexibility of the kind found in machine glazed products.[36]

Of necessity the main emphasis was upon cutting costs and eliminating loss making operations. A war on waste was conducted under the strict paternal eyes of Colonel Davidson, fire insurance was kept to a prudent minimum, and efficiency bonus payments made their appearance in the accounts, soon exceeding £2,000 a year.[37]

The most notable attempt to prune loss making obligations was the sale of the subsidiary firm the Colonial Paper Company Limited, established towards the end of the 19th century to handle Davidsons business in New South Wales. The company had stagnated since the end of the war, its assets had shrunk in value from a pre-war total of more than £16,000 to less than £12,000 in May 1928, of which £3,300 represented goodwill, and over the immediate years before the sale from 1922 to 1927 a small net loss was recorded by the parent company. The sale realised £5,854 after deducting some £600 for various expenses.[38]

In keeping with the search for greater efficiency the annual company accounts from 1928 onwards were presented in a clearer manner. The profit and loss account distinguished trading profit for the year from other sources of income and on the debit side directors fees and interest payments on loans other than debenture interest, which had always been stated, were openly revealed.

Examining the accounts in the new format it is clear that in the three years 1928–30 a modest profit ranging from £3,351 to £5,081 was made on trading operations whilst interest and dividends from investments produced some £200 a year in addition. However when interest of more than £4,800 a year on the debentures and bank loans had been paid there was an overall deficit in 1929 and 1930. In 1928 there was a modest surplus of £468, but when allowance was made for writing down the value of stocks due to falling prices, for the reduced value of certain investments held by the company, and for the loss on the sale of the Colonial Paper Company Limited, this surplus was converted into a loss of nearly £11,000.

The overall deficits of £1,427 in July 1929 and £657 in July 1930 were the lowest recorded since 1922, but any hopes that the improvement would continue were destroyed by the removal of the import duty on wrapping papers for six months from May 1931 and the growing impact of the world slump. The paper industry suffered less unemployment than British industries generally but the percentage of workers in the paper industry who were unemployed had reached 10 per cent by November 1930 and then rose rapidly to nearly 18 per cent in September 1931 before falling again to 10 per cent by the end of the year.[39]

At Mugiemoss the sudden exposure to the full force of foreign competition, plus declining demand for wrapping papers as a result of the depression, once more threatened disaster. In the year ending 31 July 1931 there was a loss on trading operations of more than £10,000 and an overall deficit after meeting interest charges of £14,425.

The directors decided to take drastic action even before the full results of the financial year were to hand. On 7th July 1931 shareholders were invited to attend an Extra-ordinary General Meeting in order to approve some major changes and accordingly to adopt fresh Articles of Association for the company in place of the existing ones.[40]

The problems facing the company were summarised clearly in two of the opening paragraphs of the letter sent to shareholders:

> 'As you have no doubt surmised, the Company has lately been coming through a very difficult time. Up to eighteen months ago, when the world slump began, steady improvement had been made for some years, but the general slump retarded this progress, and now the cancellation of the Safeguarding Duty on wrapping paper has further added to the difficulties of the situation'.
>
> 'Generally speaking, wrapping paper can now only be made and sold in this country below the cost of production, and, if this Company were dependent solely on the results of making wrapping paper, the outlook at the moment would be well-nigh hopeless. It so happens, however, that paper-making is only one part of the Company's business. Its other activities at Mugiemoss consist in Bag-making, both by hand and by machine, Carton-making, Roofing, Felt-making, Printing and Lithographing, and all these activities can be carried on profitably, without paper-

making, by purchasing the required paper from other sources. There are also other directions in which profit can be made by purchasing instead of manufacturing the paper required.'

The directors therefore proposed to curtail the manufacture of paper until changed conditions, such as the re-imposition of import duties, again made it possible to make a profit. Regrettably this would mean the loss of some 100 to 150 jobs. Bag making and various other activities at the mills would be continued mainly using paper purchased from other firms.

The sales organisation of the firm, which for many years had bought the products of other paper mills for resale and had operated as wholesale stationers as well as distributing Davidsons' own products, was now to be divorced completely from the manufacturing operations. The London office which was the most important sales office, had experienced considerable difficulties in making purchases because of delays in paying for the goods and it was felt that additional bank credit was essential and would not be forthcoming unless the sales organisation was separated from the heavily indebted existing Company. Davidsons Paper Sales Limited, a private company, had therefore been formed as a wholly owned subsidiary in June 1931 to carry on the selling part of the business.[41]

As a condition for accepting these changes the North of Scotland Bank, which already held 20,000 partly paid £1 shares of the 1881 issue as security, stipulated that the outstanding balance of 10 shillings per share should be called up. Accordingly holders of these shares were informed by letter on 6 August 1931 that the remaining 10 shillings per share had to be paid to the Bank in two equal instalments on 31st August and 16th October.[42]

Davidsons' shareholders duly approved the changes and within months of the decision external economic and political changes once more swung in favour of the Directors' fight to save the firm. Import duties on wrapping and packing papers were re-imposed in November 1931 at the much higher level of 50 per cent *ad valorem*. Subsequently the duties were reduced to 20 per cent in April 1932 and after being increased to 25 per cent in September the signing of commercial treaties with Norway and Sweden, which were the main suppliers of wrapping papers for the British market, led to a further reduction of import duties on wrapping papers purchased from these sources to sixteen and two thirds per cent.[43]

The directors responded swiftly to the improved climate. Within days of the re-imposition of import duties paper production at Mugiemoss had returned to the levels recorded in the earlier months of 1931. Investment in new plant, which had sunk to the derisory figure of £713 in 1931, was now resumed on a serious scale and it can be seen from Table 17 that over the three years 1932–34 more than £13,000 was spent on new plant at Mugiemoss, including nearly £2,400 invested in improved systems of accountancy and production control.[44]

The more optimistic outlook was reflected only partially in the firm's financial performance. The parent firm, divorced of its selling operation, made a profit on its operations, even after finding the capital for investment in new plant in each of the years 1932–34. When interest on the debentures and bank loans had been paid there was still a surplus, ranging from more than £4,400 in 1932 to £741 in 1934. Ironically the overall picture was spoilt by the weak performance of the sales organisation, held up to shareholders in 1931 as one of the profitable sectors of the business. The newly formed subsidiary, after recording a modest profit of £31 in 1932, made losses of £2,877 in 1933 and £3,393 in 1934 thus producing an overall deficit for the parent firm. The cumulative deficit on the Profit and Loss Account increased correspondingly, reaching a record total of £46,648 in 1934.[45]

The time had clearly arrived for more drastic measures. New products yielding greater profit margins were needed, but above all reform of the capital structure of the company was essential to reduce the burden of interest payments and to remove the insidious threat to the very existence of the firm posed by the mountain of debt.

CAPITAL RECONSTRUCTION, 1935

Capital reconstruction of the company had been one of the stated aims of the directors for several years. Their annual report to shareholders in 1928 spoke of submitting a scheme for consideration 'if the present improvement in trade continues throughout the year', and the circular letter sent during the crisis in July 1931 again promised reorganisation of the company's capital once the various other measures taken had proved successful.

However the decision was finally precipitated not after a run of good results when shareholders and the general public would be receptive to changes in the capital structure but as the result of another financial crisis. By the early summer of 1935 it was obvious that the firm was heading for a considerable loss on its trading operations and therefore a substantial overall deficit on the year's activities, but the immediate problem was the increasing difficulty found in paying the firm's current bills, including Aberdeen Corporation for the supply of electricity. On 30 May 1935 therefore the directors sent a circular letter to the firm's creditors summoning them to a meeting on 10 June.

The financial position, quite apart from cash flow problems, was indeed depressing. The firm owed some £100,000. Of this sum more than £30,000 was owed to trade creditors, £12,000 was in the form of temporary loans, and the balance of more than £57,000 was owed to the North of Scotland Bank in two separate accounts.[46]

The position of the shareholders and debenture holders was scarcely more encouraging. The debentures were not secured by a mortgage or any other charge on the company's assets nor were they entitled to preferential treatment if the company was wound up. The shareholders had received no dividend since 1921. On the stock exchange the valuation placed on both shares and debentures was a fraction of their nominal value. For some months the £100 debentures had been quoted at £28 or less and the £1 shares at no more than 1s 3d.[47]

At the creditors' meeting the directors sought to disarm critics by stressing that their object was not simply to save the existing business but to take a bold new initiative, namely the manufacture of waterproof paper and board employing a new process patented by International Bitumen Emulsions Limited, who had granted Davidsons a licence to use the patents in the United Kingdom. Fresh capital would be needed to bring the process to full production but the new products would enable the firm to operate on a profitable basis. The first essential, however, was the reconstruction of the company and with this in mind the directors submitted a memorandum for consideration.[48]

After questions, discussion, and further explanation by the directors the meeting passed three resolutions.

1. That this meeting is of the opinion that a Scheme of

Reconstruction is desirable and that a Committee of Creditors be appointed to consider whether a Scheme of Reconstruction is practicable, and if so, to co-operate with the Directors in the preparation of a formal Scheme of Reconstruction to be submitted for sanction of the Court.

2. That a Moratorium in respect of debts incurred prior to 31st May 1935 be granted for the period necessary to obtain the sanction of the Court to the Scheme, and that the Company be authorised to pay running expenses and the expenses incidental to the preparation and completion of the Scheme.

The third resolution appointed the Committee of Creditors. This consisted of seven men, five representing the trade creditors, plus G.A. Williamson of the Third Scottish Northern Investment Trust, Aberdeen, a debenture holder, and James Mearns of Aberdeen, a loan creditor, who was A.T. Dawson's fellow director in Morrison's Economic Stores Limited.[49] The trade creditors – D. Hamlyn, A. Pyper, Bramhall, S. Nicol and T. Hutchinson – were drawn from the larger firms, representing respectively two paper manufacturers, Inveresk Paper Company Limited, London and William Collins and Company Limited, Glasgow, T.Y. Nuttall Limited, Manchester, and Berner Nicol and Company Limited of London, pulp merchants, and the Forth and Clyde Coal Company Limited, Glasgow.

The co-operation of the North of Scotland Bank was secured by an agreement over the £37,130 owing on their No.1 Account. The Bank regarded the overdraft of £20,000 on their No.3 Account as adequately safeguarded by the existing security held. The debt on the No.1 Account was to be reduced by the Bank taking over the company's heritable properties already held as security, namely Shore Brae and Shiprow Aberdeen and the warehouse in Frederick Street Glasgow together with £3,800 of the firm's debentures, also held in security. This would reduce the debt by £3,000 and £1,267 respectively (valuing the debentures at 6s 8d per £1), thus leaving the sum of £32,863 outstanding.

This sum was unsecured and the Bank was reluctant initially to accept any new debentures to be created under the Scheme of Arrangement in lieu of the debt. However Peter Dawson of Drumcoille, Braco, Perthshire – A.T. Dawson's father – agreed to purchase the debt at an agreed figure should the proposed scheme be sanctioned.[50]

The Committee of Creditors met on 19th June. It was presented with information about the new manufacturing process, a statement prepared by the company's auditors and a memorandum from the company's lawyers, and it was noted that the directors had received assurances that at least £20,000 of new capital would be forthcoming if the Scheme of Arrangement went ahead. The Committee then resolved that the reconstruction of the company was practicable and should proceed.

Separate meetings of the shareholders, debenture holders and the various groups of creditors held on 30th September approved the Scheme. The Court of Session added its seal of approval, giving final sanction on 28th November.[51]

Under the Scheme the 97,886 fully paid £1 ordinary shares were reduced in value to one shilling per share and 40 shares on which calls were outstanding were cancelled, as were 2,074 unissued shares of £1. However an additional 1,902,114 shares of one shilling each were created thus raising the authorised share capital again to £100,000.

Two new classes of debentures were created. The 5 per cent convertible mortgage debentures were convertible into ordinary shares of the company at par at any time within five years from 1 June 1935 or redeemable in cash on 1 June 1940. The four and a half per cent income debentures were repayable at par at the end of 15 years from 1 June 1935 and the interest was non cumulative, payable out of the current year's net profits after making provision for depreciation and anticipated losses or other contingencies approved by the company's auditors.

Creditors with debts of less than £10 were repaid in full in cash. The holders of the existing debentures and the more substantial creditors were given the choice of converting their claims against the company either into the 5 per cent debentures at the rate of 6s 8d per £1 (one third) or into four and a half per cent debentures at the rate of 20s per £1, but they were permitted to combine the two options, reclaiming part of their debt by one method and the remainder by the other.

Clearly the choice required a nice balance of judgement. Repayment of one third of the debt after five years, with the virtual certainty of 5 per cent interest per year until then and the option to convert the holding into ordinary shares had to be set against repayment of the whole debt after fifteen years plus a considerable risk that in at least some of these years the interest of four and a half per cent on the sum would not be paid.

Sufficient debentures of both classes were created to satisfy the applications from creditors and the existing debenture holders. In addition further 5 per cent convertible mortgage debentures were created to raise fresh capital, at least £20,000 coming from persons who had promised the directors they would do so if the Scheme of Arrangement went ahead, and up to £15,000 from the shareholders in general.

The North of Scotland Bank acted as trustees for both classes of debenture holders with the right to appoint three directors to the Board of the company, nominated at a joint meeting of debenture holders each year. This right ceased five years after 1st June 1935, but if a large number of the mortgage debentures were converted into ordinary shares or were bought up by the company before June 1940 the number of directors appointed was to be reduced at the rate of one director for each £20,000 of debentures so converted or cancelled.

The completion of the Scheme of Arrangement marked a watershed in the firm's history. The three new directors appointed under the Scheme, increasing the number of directors to six, symbolised the break with the past. Only two members of the Board were now drawn from the Davidson family and the three newcomers brought with them substantial top level experience of occupations unrelated to papermaking.

R.W. McCrone of Charlestown, Fife, was already a director of The British Oxygen Company Limited and of Metal Industries Limited. J.A. Montgomerie for some years had been Managing Director of Montgomerie, Stobo and Company Limited, Victory Oil and Colour Works, Glasgow and later also became a Director of British Bitumen Emulsions Limited and International Bitumen Emulsions Limited, the firm which had licensed Davidsons to use the new patent waterproofing process for paper and board manufacture.[52]

J.C. Duffus was a member of the legal firm of Wilsone & Duffus Advocates, based at 7 Golden Square, Aberdeen. He stayed on after the war as Chairman of the Board and subsequently played a vital part in the post war programme of expansion.[53]

The capital of the company was now completely transformed. The authorised capital consisted of £100,000 in ordinary shares of one shilling, £66,500 in 5 per cent convertible mortgage debentures of £1 and £20,000 in four and a half per cent income debentures also of £1 denomination. By 31 July 1936 when the next financial year ended a large proportion of the debentures with a nominal value of more than £85,000 had been issued together with shares to the value of £4,894. An overdraft of £20,000 on the No.3 account with the North of Scotland Bank was the sole reminder of the mountain of debt and there was a Capital Reserve of £44,033 to finance new developments and to act as a cushion against possible losses until the new products were in full scale production.[54]

NEW DIRECTIONS, 1935–1939

The first of the new developments, the manufacture of waterproof paper by the patent Ibeco process, was soon put into operation. The firm, like some of its competitors, had been producing waterproof papers and paper felts by coating their surface with tar since the late 19th century. However the surface of these papers and felts was liable to deteriorate particularly under extremes of temperature, either hardening and cracking in the cold or sweating out in the heat. The Ibeco process avoided this problem since the bitumen was added to the pulp in the beater in the form of an emulsion before the pulp was run onto the Papermaking Machine, thus making the waterproofing an integral part of the paper fibres. Ibeco Kraft combined the lightness and strength of all Kraft papers with complete imperviousness to moisture and this quality was retained even when the paper was folded or roughly handled.[55]

The new product was intended principally for the wrapping paper market, particularly where goods in transit or lengthy storage required protection from the damp. However its versatility led to its use in a variety of applications. One of the earliest major outlets was in the construction of concrete roads and other concrete products. Roads, floors and pre-cast concrete products such as slabs and kerbstones made from ordinary Portland cement were not ready for the stress of daily usage until an interval of 2–4 weeks after the concrete had been made and even when rapid hardening Portland cement was introduced subsequently the interval was still a minimum of six days.[56]

During the 'curing' period the concrete hardened gradually and attained its maximum strength but cracks would appear if there was undue loss of moisture. It was important therefore to protect the surface of the concrete and in the case of a road the underneath side also from drying out quickly under the influence of wind and high temperatures or through contact with dry stones, clinker and subsoil. In cold weather also protection was needed against frost damage. Sand, soil, sacking, tarpaulins and wooden shuttering provided alternative ways of protecting the surface of concrete roads and floors but waterproof paper increasingly came to be used as more convenient and economic and as an underlay to protect the bottom surface it was especially useful.[57]

Orders for Ibeco paper from the construction industry thus swelled demand from customers requiring wrapping paper. The original East Lancashire Road and Wembley Ice Rink were among the early projects supplied and by 1939 the firm's advertisement in the *Concrete Year Book* was able to claim that Ibeco Waterproof Kraft . . . 'is used extensively on Navy, Army and Air Force contracts, and by Corporations, Councils and Highway Constructors throughout Great Britain and Ireland'. Davidson's Paper Sales Limited was in charge of distribution but the advertisement also listed agents in Dublin and Belfast as well as Leeds, Slough and Liverpool. Advertisements were also placed in three other leading publications read by building contractors and municipal engineers.[58]

The Ibeco trademark for the new waterproof paper and board was officially registered to provide nationwide protection on 30th August 1935.[59] Production of Ibeco Kraft paper had begun on the 9th of August and by the end of the month more than 16 tons had been produced. Thereafter output increased quite rapidly as can be seen from Table 18. Nearly 71 tons of Ibeco Kraft were produced in the last five months of 1935, 301 tons in 1936, and by 1938 production had more than doubled again, reaching 657 tons. In the first six months of 1939, before the needs of war altered the pattern of demand, the upward trend continued with an output of 434 tons. The five fold increase in the rate of annual output was reflected in the growing share of the firm's production of paper, rising each year, from 4 per cent in 1935 to 19 per cent in 1938 and 25 per cent in the first half of 1939.

The other new venture, the production of millboard, required substantial investment in extra plant and because it was a new field

Table 18: Ibeco Paper Production, 1935–1939

Year	Total Output of Paper (Tons)	Ibeco Output	
		Tons	Per cent
1935[1]	1960	71	3.62
1936	3723	301	8.08
1937	5257	573	10.90
1938	3456	657	19.01
1939 First half	1759	434	24.67
Second half	2116	663	31.33

Source: Daily Report Books, 1935–1939
Note: 1 Covers the period 4 July–31 December 1935.

for the firm there was a longer interval between initiating the new project and the attainment of the planned level of production. The existing Papermaking machines at Mugiemoss were quite capable of producing thinner kinds of paper board such as *cardboard*, *pasteboard* made by pasting good quality paper onto both sides of the inferior *middle*, and *ivory boards* formed by pressing two or more sheets of fine white paper together.

The thicker millboard however was made by a somewhat different process. The typical Board Machine in use at the beginning of the 20th century had as its main feature a large hollow wire-covered cylindrical drum onto which the pulp flowed direct from the stuff chest. The pulp adhered to the outer surface of the drum as it revolved and the wet sheet of pulp was then deposited on an endless felt which carried the sheet between two couch rolls at the other end of the machine. The sheet of pulp was wound continuously round the upper roll until it reached the required thickness when the wet board was removed and taken away for pressing, to remove excess moisture and to 'close up' the sheet into a hard solid board, and then for drying and glazing.[60]

The decision to embark upon millboard production extended the firm's operations into supplying the rapidly growing market for boxes, cartons, and containers and also for cheap panels of various kinds including motor car interiors and the backs of wireless sets. Foreign manufacturers with ready access to cheap supplies of wood pulp dominated the British market for millboard. Yet British producers, drawing on supplies of waste paper collected in Britain for a large part of their fibrous raw materials, had begun to compete on a more equal footing, doubling their output between 1924 and 1930.[61]

The introduction of a 15 per cent *ad valorem* duty on imports of board weighing more than 90 pounds per ream in April 1932 and its increase to 20 per cent in May 1934 gave a further stimulus to expansion.[62] The most spectacular developments took place in Southern England, notably at St. Anne's Board Mill Company Limited of Bristol, the Thames Board Mill at Purfleet, Essex and Jackson's Millboard and Paper Company Limited at Bourne End, Buckinghamshire. In each location large board machines more than 100 inches wide were installed in newly built or extended premises, substantially increasing productive capacity.[63]

The virtual absence of similar developments in Scotland, where for example, William Martin of Ayton, Berwickshire with a relatively narrow Board Machine 58 inches wide was the sole millboard producer listed in the *Directory of Papermakers of the United Kingdom* in 1935, might have daunted the Mugiemoss directors. However the lack of major competition within Scotland also provided an opportunity and may well have been an important consideration when the directors examined their future strategy during the 1935 crisis.

The directors recruited for the new venture Frank Williamson, an experienced board maker. A wide Fourdrinier paper-making machine previously used to make newsprint was purchased second hand from Hendon Paper Works Company Limited of Sunderland. The dry end was retained and a new wet end made to Davidson's own design was ordered from Masson Scott. The machine when assembled was designed to produce board more than 12 feet (144 inches) wide, making it one of the widest of its kind in the UK.[64]

A delay of some four months in supplying the new parts meant that production did not begin until August 1936. Even then production increased rather slowly. Only 286 tons of board had been made by the end of the year and it was the summer of 1937 before the teething problems had been fully overcome and the machine was operating at the rate of output originally envisaged, producing 1969 tons in the second quarter of the year and more than 2,200 tons in the following quarter. A recession in the paper and board industry in the first half of 1938 led to short time working and cut production over that period to less than 1900 tons and although there was a strong recovery, taking output to a record 2743 tons in the final quarter of the year, total production for 1938 was nearly 400 tons less than in 1937.[65]

In 1939 in the last months of peace, however, further expansion was very evident. Output exceeded 5,400 tons in the first half of the year and the figure for the second quarter, 2831 tons, broke all previous records.

The rapid growth of board production coincided with the formation of the Northern Wastepaper Company Limited in May 1937. The three directors – John Lawrie, a metal merchant, William Munro Cattenach, a solicitor, and David Watson, a manufacturer's agent – were Aberdeen men and the head office was located in Wellington Road near Aberdeen harbour. There is little doubt that the firm was supplying the board making plant at Mugiemoss, although Davidsons also placed substantial orders for waste paper with Jebb Brothers Ltd of Glasgow and John Leng & Co Ltd of Dundee, and closer links were forged in October 1937 when Frank Williamson, the mill manager, became a director. In March 1939 the connection was taken a stage further and the head office was switched to Mugiemoss.[66]

The principal product of the board mill in the first 18 months of operations was natural chipboard, millboard without any special treatment to emboss the surface or give it a coloured lining, which comprised some 90 per cent of production. The depression in the early months of 1938 stimulated experiments to produce a more varied product and a search for new markets which resulted in a much more diverse output. Natural chipboard accounted for little more than 27 per cent of production in 1938. Chipboard with a top ply of brown, grey, white or various other colours constituted nearly 10 per cent of the total, reflecting the growing demand for colourful packaging as a means of attracting consumers.[67]

Embossed boards in various colours accounted for a further 23 per cent of production and leather boards, made with pulp containing some leather chippings and used for bookbinding, covers, and the insoles and heels of boots, for nearly 5 per cent of the total. Ironically Ibeco Board was a very minor item.

However the most significant addition to the range was the production of liners for plasterboard. Plasterboard, which consists of a core of plaster of Paris sandwiched between and bonded to two sheets or *liners* of millboard, was a cheaper substitute for the laths and plaster used in making ceilings and interior walls of houses.

British Plaster Board Limited had begun making plasterboard at Wallasey in 1919 but for several years only *baseboard* was produced, which required one and preferably two coats of plaster applying *in situ*. However the introduction in the 1930s of an improved product, *wallboard*, a self finished material to which wallpaper or other decoration could be applied direct or more usually after applying a thin skim of plaster to conceal and seal the joints between the boards, increased the economies possible from its use. As the building boom in England gathered force in the early 1930s speculative builders of small houses seized the chance to cut costs, although wallboards were still used largely for ceilings, and demand soared. British Plaster Board Limited erected a second factory at Erith in Kent and a third one at Cocklakes in Westmorland. Meanwhile competitors had appeared on the scene, principally Gyproc Products Limited, an Anglo-Canadian venture, which started a factory at Rochester in Kent and when this was operating at full capacity in 1936 opened another factory in Glasgow.[68]

The board machine at Mugiemoss was soon supplying both the leading Plasterboard producers. The first order for plasterboard liner from the Scottish Gyproc factory was obtained in 1937. An order for 4,000 tons from British Plaster Board Ltd followed in June 1938 and a year later nearly half of this had been delivered in monthly instalments ranging from 290 tons in July 1938 to 11 tons in February 1939. Over 1938 as a whole the combined output of liners for BPB boards and Gyproc boards accounted for more than 26 per cent of the output of the board machine.[69]

Measured in terms of tonnage produced board-making had now become a major part of the Davidson enterprise. In 1937 board production accounted for 58 per cent of the combined output of paper and board and this figure increased to 66 per cent in the following year and to no less than 75 per cent in the first half of 1939.[70]

It was some time before the new developments came to fruition financially. The modest trading profit of £718 earned in the financial year ending 31 July 1936, which became an overall deficit of £2,344 after payment of interest on loans and debentures, clearly owed virtually nothing to the production of board or Ibeco paper. In the following year the board machine, starting in August, had scarcely begun production on a very limited scale when the three months' strike commencing on 3rd October drastically curtailed all output of paper and board. There was a deficit on trading, after meeting all expenses of £2,598, giving a total loss of £6,804. Moreover an additional sum of £10,052 was taken out of the Capital Reserve and written off as development costs for the machine.[71]

In the financial year 1937–38 a severe recession in the U.K. board industry in the second half of the year was reflected in reduced production at Mugeimoss. The output of paper was also depressed and a serious fire in March 1938 which destroyed practically all stocks of raw materials and some finished goods with a total value of more than £7,000, although covered by insurance, added to the problems. There was a deficit of £10,631 on trading operations after meeting all expenses, and overall, when interest payments had been covered, a substantial loss of £14,691.[72]

It was not until the last year of peace when the production of board and Ibeco paper were at record levels that the benefits of the new initiatives were really evident. Despite a decline in the price of Kraft paper and a consequent reduction in profit margins the balance sheet in July 1939 revealed a trading profit of £6,744 and there was an overall surplus for the first time since 1932 of £3,045. Sensibly the directors resisted the temptation to declare a dividend but at least it must have seemed that the firm's fortunes had turned the corner and that the long dividend famine was nearly at an end.[73]

Chapter 4

The War Years, 1939–1945

NATIONAL NEEDS

When war with Germany began on 3rd September 1939 the government, with the experience of the 1914–1918 conflict in mind, had already made plans to control the nation's productive resources in the interests of the war effort. Wide ranging emergency powers were granted to the government in August, new Ministries, including Supply, Information, Economic Warfare, Food and Shipping were created, and soon a comprehensive network of boards and committees sprang into existence to issue detailed orders and directives.[1]

A system of reserved occupations was introduced by the Ministry of Labour and National Service to ensure that there would be no sudden major loss of skilled workers to the Armed Forces as in the First World War. In the Paper and Board industry at the start of the war the ages at which exemption from service in the Forces began ranged from 25 to 30 for men whilst women were exempt irrespective of age.[2]

The first Control of Paper Order was issued by the Minister of Supply on 2nd September, a Paper Control Office was established in Reading, and A. Ralph Reed, chairman of Albert E. Reed Limited, the well known paper making firm, was appointed Paper Controller.[3]

The magnitude of the task facing him can be seen in part from the figures of paper production and consumption in the last year of peace. Some 3,250,000 tons of paper were consumed in the United Kingdom in 1938, of which approximately 1,070,000 tons were imported. The production of the remaining 2,180,000 tons, plus 175,000 tons for export, was heavily dependent on imported raw materials, requiring 1,600,000 tons of wood pulp, 340,000 tons of pulp wood for conversion into wood pulp, and 311,000 tons of esparto grass, in addition to waste paper and rags obtained mainly from the United Kingdom.[4]

It seemed very unlikely that a similar level of imports could be maintained during wartime. On the other hand the demand for boards and wrapping papers would increase substantially both to meet specific needs of the Armed Forces and because of the increased substitution of containers of paper and board for wooden cases and tin cans.

The first concern of Paper Control, however, was to prevent the very marked inflation which had occured during the First World War. Maximum prices were fixed for all classes of paper, initially at the level ruling on the eve of the war, a task which Reed observed was 'the first time in the history of the trade that current market prices were recorded for every possible item of the innumerable varieties of paper'. Maximum prices were similarly fixed for the principal home produced raw materials, waste paper and rags, and for imported wood pulp and esparto.[5]

A list of merchants authorised to deal in waste paper-making materials was published and purchases of supplies from other sources was limited to the pre-war level. The Ministry of Supply from the end of October took control of all existing pre-war supplies of wood pulp, esparto, and pulp wood plus all future imported supplies due to arrive under the paper mills' own contracts and became the sole purchaser of imported raw materials.[6]

Rationing of paper supplies was delayed until the beginning of March 1940. It was imposed not because of increasing enemy action, which had remained at a negligible level in W. Europe during the first six months of 'phoney war', permitting imports of wood pulp from Norway and Sweden, the principal suppliers, to exceed the pre-war figure, but because of the need to conserve shipping space and foreign currency reserves for other purposes.[7]

Control of Paper (No.8) Order limited the quantity of paper which could be supplied by any paper mill to any particular customer to 60 per cent of the amount supplied in the corresponding period before the war, unless additional quantities were permitted under licence from Paper Control. The first licensing and rationing period ran from 3 March to 31 May 1940 and was followed by further periods of similar length. At the same time all imports of paper were forbidden except under licence.[8]

The German conquest of Denmark and Norway in April, followed by the collapse of Belgium, Holland, and France in May and June 1940 finally shattered any illusions that there might be a limited conflict ending with a British victory over a Germany seriously weakened by an economic blockade.[9] For the paper industry the immediate consequence was the loss of the major sources of its fibrous raw materials – French North Africa, Scandinavia and the Baltic – which together accounted for some 90 per cent of the industry's pre war imported supplies. The alternative source of supply from abroad was North American wood pulp which involved a longer and slower voyage, now made more hazardous as a consequence of the newly acquired German bases for bombers and submarines stretching for more than 2000 miles along the Atlantic seaboard.[10]

In this new situation the rationing system was applied more rigorously and re-inforced by various other measures, culminating in the drastic restrictions imposed by Order No. 36 in November 1941. The manufacture of such inessential products as paper serviettes, handkerchiefs, cups, plates and saucers, decorations, confetti, christmas cards, picture postcards and hair sachets for permanent waves was strictly forbidden. Maximum sizes were laid down for a variety of products ranging from labels to posters, a maximum weight was laid down for Calendars, reduced in 1943 to 2 ounces, and the number of posters displayed to advertise any entertainment in cinema, theatre, dance hall or sports arena was limited to ten for a particular event. Retailers were not permitted to wrap or pack any goods they sold over the counter except food stuffs.[11]

There was also exhortation to promote a variety of other paper saving practises. Letters were to be typed in single spacing with narrow margins on both sides of the sheet and envelopes were to be re-used several times with the aid of the ubiquitous economy label.[12]

Some major institutions gave a lead. The Post Office saved 500 tons of paper a year by extending the life of a telephone directory from twelve to eighteen months, telephone bills were sent out every six months instead of quarterly, and internal memoranda were written on scrap paper even when addressed to the highest level. The LMS Railway Company was stated to reclaim 30,000 old envelopes a week at its reclamation depots where teams of girls covered up all writing with strips of gummed paper. Even the *Police Chronicle* made its contribution, urging the elimination of jargon since simpler, fewer words would use less paper.[13]

Waste paper supplies were increased by every possible means. Local Authorities of more than 10,000 people were compelled to collect waste regularly and contests were held with prizes going to the Authority which collected the highest amount of waste paper per head. 'A mile of books' was a common target and some boroughs collected more than a million books during the span of a contest. Children were encouraged to participate and Aberdeen was praised in a leading Paper Trade Journal for a scheme which rewarded the leading collectors with a free visit to the cinema and a trip to a local paper mill where they could see waste paper being pulped. Householders were exhorted that only greasy or dirty paper should be used to light fires and from March 1942 it became a punishable offence to burn, destroy or throw away waste paper.[14]

Home grown supplies of straw were also sought as an alternative fibrous material. The Papermakers Association held discussions with the National Farmers' Union, the Hay Traders' Association and Paper Control, and in December 1940 the Papermakers Straw Trading Company Limited was formed with directors drawn from various papermaking firms to purchase straw centrally for use in paper and board production.[15]

Throughout the war years government intervention relied heavily on the co-operation and assistance of the paper industry both in individual firms and collectively through the Papermakers' Association.[16] The relationship was most severely tested when the Board of Trade drew up plans, in line with those made for some other industries, to concentrate production in certain mills thus releasing factory space and manpower elsewhere for war purposes. By January 1942 176 paper mills had been granted nucleus status, indicating that their production must be maintained for the war effort, 12 mills had already closed, and 51 other mills were at work, including 22 which were due to lose their nucleus certificates at the end of the month. The Papermakers' Association, whose members feared that mills which ceased paper production might be lost permanently to the industry, argued that the government's aims could be realised just as effectively if the industry voluntarily released more factory space and manpower in addition to the substantial contribution already made in this direction. After lengthy discussions this view was accepted, an agreement on releasing further labour and factory space was arrived at, and non nucleus mills remained in production although operating well below their normal level.[17]

For the industry as a whole as the shortage of raw materials increased despite all efforts, output declined from 2,631,000 tons of paper and board in 1939 to 1,794,000 tons in 1940 and then remained well below pre-war levels. Home grown straw supplied 141,000 tons of raw material in 1941 and more than 300,000 tons by 1943 whilst supplies of rags, ropes and similar waste fibres were maintained at some 100,000 tons a year. Waste paper consumption, exceeding 800,000 tons in 1941 and 1942, was some 200,000 tons greater than in 1939 and perhaps twice the normal pre-war level.[18]

But the drastic reduction in imports of wood pulp more than offset the increased use of home produced raw materials. By 1941 supplies had fallen to less than a third of the 1938 total of 1,600,050 tons. Subsequently U boat successes in the Atlantic and the extra demands on shipping space for the invasion of French North Africa, then Sicily and Italy, together with the build up of men and munitions from the U.S.A. in preparation for the invasion of Normandy, reduced imports still further. In 1943 consumption of wood pulp reached its lowest level at 320,000 tons. Correspondingly the output of paper and board fell to its nadir in the same year, declining to less than 1,200,000 tons, a figure which was 50 per cent of the output in 1938 and a mere 44 per cent of the 1939 output.[19]

The worst hit sector of the industry was the production of newsprint. The joint pressure of Canadian newsprint producers through their government and British newspaper proprietors, who were deeply suspicious of Bowater's dominance of UK newsprint production, ensured that Canadian newsprint was given preferential treatment over Canadian wood pulp in shipping supplies to Britain and in 1943 production of British newsprint declined to scarcely 16 per cent of the 1939 total.[20]

On the other hand board mills received very favourable treatment. The *Papermaker* reported in January 1942 that in drawing up plans for the concentration of production in nucleus mills the Board of Trade began by deciding that board mills were absolutely essential 'and in the opinion of the Government, whatever happened to any other mills, they must be kept going'. Hence even at their lowest point, in 1943, the output of boards was approximately 73 per cent of the figure for 1939 and 94 per cent of the total for 1938, a year when board production had been at a low level for several months.[21]

Production of packing and wrapping papers was also maintained at a substantial level, declining in 1943 to 63 per cent of the 1939 output. By contrast the comparable figure for printing and writing papers was only 35 per cent.[22]

THE MUGIEMOSS CONTRIBUTION

At Mugiemoss the new developments which had begun to prove their worth by 1939 were equally valuable in wartime. In combination with the long established business in packing and wrapping papers they placed the firm squarely within the two sectors of the paper industry most able to benefit from the changed conditions.

Yet there was no room for complacency. Customers' specifications had to be met within the constraints imposed by the availability of raw materials and by government controls over prices, raw material contents and qualities. Designation as a nucleus mill in May 1941 removed some of the anxieties for the future of the plant but throughout the war ingenuity and adaptability were essential qualities in maintaining production at a high level.[23]

At the outset of the war concern over raw material supplies led to closer links with the Northern Wastepaper Company Limited which became an officially authorised dealer in waste paper-making materials in September 1939. A.T. Dawson joined the Board of Directors, the share capital was increased, and Dawson and Williamson, the mill manager who was already a director, acquired more shares, giving them a controlling interest.[24]

In the early months of 1941 a serious shortage of materials developed as mills making printing and other papers competed fiercely for supplies of the cheapest waste papers, Grade No.13, which was an established major source for Mugiemoss. On 29th January Colonel Davidson informed the London Office that

> The waste paper situation gets worse every day, and we are at our

> wits' end here to keep the machines running. Our stocks of waste paper are completely exhausted, and we are just working from hand to mouth.

On 2nd April he told the Edinburgh office

> We are still having great trouble in getting enough waste paper, and the Board Machine often stands for a few hours in the morning, waiting for the first load of waste to come down from the station.[25]

Tighter government controls soon eased the situation. Group 13 waste became subject to licence and supplies were allocated almost exclusively for the production of boards. The directors' annual report in 1942 referred to a considerable improvement in the supply of raw materials and noted that the Board and Paper Mills had been on full production during the past year from 1st August 1941 to 31 July 1942.[26]

Direct enemy action had relatively little effect on the volume of production. The Mugiemoss plant escaped without damage despite several air raids on the city of Aberdeen which destroyed life and property. The Glasgow warehouse of Davidsons' Paper Sales Limited was completely demolished by a bomb in September 1940 whilst the Liverpool and London warehouses had narrow escapes, an incendiary bomb in the latter falling harmlessly to the bottom of the lift shaft. However perhaps more disruption was caused in the autumn of 1940 by the temporary closure of railway stations in London and Southern England generally, which led to considerable delays in the delivery of finished products.[27]

The composition of output meanwhile changed greatly to meet wartime needs. Imitation Krafts were introduced in May 1940 and by late August Colonel Davidson was informing Paper Control that Imitation Krafts, which were made of 100 per cent home produced raw materials, now constituted the main line manufactured in the Paper Mill and that we have 'got our customers regularly established on this quality.' By the end of the year even the machine glazed sulphite paper used in bag making contained no more than 50 per cent of sulphite pulp mixed with inferior materials and the term 'Imitation' might have been applied to the whole range of the Mill's papers.[28]

The Board Mill also concentrated on plainer and more utilitarian products. The production of embossed boards and chipboard with top plies of various colours dwindled to a trickle. Medium sized chipboard became the main product, accounting for 28 per cent of output in 1940 whilst natural chipboard and hard sized chipboard constituted a further 20 per cent and 7 per cent respectively.[29]

The composition of plasterboard liners which accounted for nearly 26 per cent of the output of the Board Mill during the year, was dramatically adjusted to the new realities of wartime when the Paper Controller in August 1940 decreed that unlined plasterboard, given an extra coating of size to compensate for the now forbidden grey liner, should be the new standard product. Colonel Davidson writing to B.P.B. Limited, Gyproc Limited and I.C.I., expressed his regret and assured them that it was 'not the wish of the Board Makers that any change should be made'.[30]

The direct contribution of Mugiemoss to the war effort cannot be gauged, for the end usage of the majority of products leaving the mill is not known. Certainly rocket paper and cases for parachute flares were among the products made, whilst Colonel Davidson writing to the London office in February 1941 observed that the Board Mill was running nearly 100 per cent on Government work.[31]

Ibeco paper once again proved its versatility. It was supplied for aerodrome contracts in laying concrete runways as the RAF expanded. It was also used to make portable sandbags containing some 10 pounds of sand to protect the public and with the danger of small incendiary bombs in mind the Liverpool office of Davidsons' Paper Sales Limited produced a leaflet demonstrating the value of a Three Ways Sand Container.[32]

However a major use of Ibeco in the earlier months of the war, when black out regulations had been imposed to conceal from enemy bombers the lighted windows of towns, cities, and all manner of strategic targets at night, was as a cheap blackout material impervious to moisture. Woolworths' stores throughout the Midlands and Southern England sold sheets of black Ibeco together with the necessary attachments to make them into a roller blind capable of covering the normal sized domestic window. Production of black Ibeco began on a small scale in the second quarter of 1939, increased markedly from July to October, died away during the phoney war and then reached a peak, exceeding the production of the standard brown Ibeco, in the second half of 1940 when enemy bombing began in earnest.[33]

Other products made at Mugiemoss for civilian use made their contribution to ultimate victory by sustaining the health and morale of the population. Some foods still required wrapping or packing in bags and bags for sugar, flour, and tea were made for various customers.[34]

The drive to eliminate as far as possible the use of wrapping and packing papers for consumer purchases generated orders of a different kind. Paper Control when issuing licences gave favourable treatment to applications for the manufacture of paper carriers since these eliminated the need for wrapping separate items. Hence Mugiemoss received an order for 250,000 carriers from Marks & Spencer in November 1940 and one for 349,000 carriers from Woolworths the following January.[35]

The success of the managers and directors in obtaining Government contracts and attracting orders for approved civilian needs can be seen in Tables 19 and 20. The production data is missing for virtually the whole of the years 1941 and 1942 but it is clear that the output of paper in 1940 and 1944 was greater than in either 1936 or 1938 and amounted to nearly 80 per cent of the level for 1937,

Table 19: Output of Paper, Machines Nos.1–3,[1] 1935–1945

	Quarterly Output (tons)[2]				Annual Output (tons)
Year	I	II	III	IV	
1935	—	—	947	1013	1960[3]
1936	1218	1077	1127	301	3723
1937	1372	1445	1224	1216	5257
1938	1164	825	622	845	3456
1939	848	911	918	1198	3875
1940	1127	1042	928	1017	4114
1941	—	—	—	—	—
1942	—	—	—	—	—
1943	—	815	779	889	2484[4]
1944	901	999	987	1193	4080
1945	1045	1096	975	1145	4261

Source: Daily Report Books, 1935–45

Notes: 1 Machine No.1 was taken out of production on 18 April 1938.

2 The Daily Report Books covering the months before 4 July 1935 and the period 30 January 1941–16 April 1943 are missing.

3 Covers only the period 4 July–31 December 1935.

4 Covers only the period 17 April–31 December 1943.

Table 20: Output of Millboard, Machine No.4, 1936–1945

Year	Quarterly Output (tons) [1]				Annual Output (tons)
	I	II	III	IV	
1936	—	—	11	275	286
1937	1277	1969	2229	1740	7215
1938	762	1137	2212	2743	6854
1939	2594	2831	2907	2820	11152
1940	2906	3337	3397	3703	13343
1941	—	—	—	—	—
1942	—	—	—	—	—
1943	—	2487	2795	2927	8209 [2]
1944	2681	3208	3078	3329	12295
1945	2674	3222	2800	3344	12040

Source: Daily Report Books, 1935–45
Notes: 1 The Daily Report Book covering the period 30 January 1941–16 April 1943 is missing.
2 This figure is for less than 9 months, from 17 April–31 December 1943.

the most successful peacetime year. In the last three quarters of 1943 when imports of wood pulp were at a nadir and production of paper and board nationally at a correspondingly low ebb, output of paper at Mugiemoss amounted to nearly 64 per cent of the corresponding months of 1937.

The record of the Board Mill was more impressive. The highest peacetime output of 2,831 tons in the second quarter of 1939 and 2,907 tons in the third quarter was exceeded in three of the quarters of both 1940 and of 1944, the overall production in these years comfortably exceeding the 1939 total. In the more difficult circumstances of 1943 the level of output from 17 April to 31 December was only some 300 tons less than the corresponding period of 1939.

The relatively high level of production maintained during the war helped to achieve very satisfying financial results. A substantial profit, derived almost wholly from trading operations, was made in each of the six financial years from 1940 to 1945, ranging from £36,258 in 1940 to more than £81,000 in 1945. Overall profits amounted to £288,420, an average of more than £48,000 a year.[36]

The high level of taxation imposed to finance the war effort removed more than two thirds of this sum. Nevertheless the accumulated deficit of £20,795 on the profit and loss account was paid off in 1940 and £23,545 was added to the General Reserve

Fund in four consecutive years 1942–45, to finance future developments or to meet unexpected liabilities. The sum of £34,188 in the Capital Reserve Fund, meanwhile, remained intact as a further buttress for future expansion.

The shareholders were not neglected. A dividend of 6 per cent was paid in 1940 and for the remaining years of the war a dividend of 10 per cent was paid each year.

The directors, who had drawn no fees for their services since the capital reconstruction of the company in 1935, claimed £1,000 for their services over the past five years in July 1940. Fees of £350 a year were taken thereafter until 1945 when the figure was raised to £700 in line with the record profits of that year.[37]

The improved fortunes of the firm were also to be seen in its capital structure. The Court of Session in January 1941 approved the extension of the time limit for the repayment of the 5 per cent mortgage debentures or their conversion into ordinary shares to 1st June 1945, but nearly £40,000 of the debentures were in fact converted in the financial year 1942–43 and a further £12,173 the following year, leaving only £1,148 of debentures for conversion by the final date.[38] Moreover all the four and a half per cent income debentures were repaid at the end of July 1945, thus eliminating the annual burden of interest charges which had to met before a dividend could be paid on the ordinary share capital.[39]

After six years of conflict the firm thus emerged in a stronger position than in 1939, in a condition which made possible, given vision and determination at the highest level, post war expansion into new fields of activity. Such qualities were in fact not lacking. As early as 1942 the directors' annual report had assured shareholders that schemes were in preparation for post war development. In November and December 1944 when Germany's defeat seemed near, advertisements for the firm's sulphite and kraft papers and millboard products were placed in the *Papermaker* for the first time, and less than a year later when the war was scarcely over a London firm of boxmakers was acquired as the first instalment in a programme of expansion.[40]

Chapter 5

Post War Expansion and New Horizons, 1945–1953

MODERNISATION AT MUGIEMOSS

The Board of Directors which undertook an ambitious programme of development during the remaining years of the 1940s comprised a nice blend of youth and experience. Some of the older directors sought a quieter life once the war had ended. In September 1946 A.T. Dawson, who had become a director in the financial crisis of 1926 and had played an important part in the reconstruction of the firm 10 years later, decided to retire and took up residence in Jersey. R.W. McCrone and J.A. Montgomerie resigned at the same time. Later in the year at the Annual General Meeting Colonel Davidson who was now 66 retired from the posts of Chairman and joint Managing Director.[1]

The vacant seats on the Board were taken by three younger men whose ages ranged from 32 to 37 – James Mearns Dawson, who was A.T. Dawson's son, William Ranald Stewart Mellis and David Peter Davidson, Colonel Davidson's son, who was Sales Manager. He had joined the firm after completing his education and had been a director of Davidson's Paper Sales Ltd since December 1938. Mellis was managing director of Mitchell and Muil Ltd an Aberdeen firm of wholesale and retail bakers and biscuit makers as well as an old family friend.[2]

The more youthful balance was strengthened when James Partington joined the Board in July 1948 at the age of 37. He had been company secretary for four years when promoted.[3]

The other three directors had been at the heart of the firm's recovery after facing near extinction in 1935. J.C. Duffus the Aberdeen lawyer who had joined the Board in 1936 now became chairman. Frank Williamson, who had taken charge of the Board

Mill at its inception and had been a joint managing director since 1943, was made the sole managing director. Colonel Davidson remained on the Board to give the benefit of his advice and long experience.[4]

The directors embarked upon a substantial modernisation programme at Mugiemoss as soon as the war ended. However the plans were slow in coming to fruition both because of post war shortages of skilled labour, coal, steel, and building materials, as well as dollars to pay for N. American imports, and because of the retention of a network of government controls, which mitigated the worst effect of the shortages but at the same time imposed irritating bureaucratic delays.[5]

One of the first major decisions was to install a completely new boiler and power plant with four boilers, each capable of generating 32,000 lbs. of steam from low grade fuel. The first steps were taken in 1945 but it was the summer of 1948 before the work was finally completed.

In the same programme extensions were made to the Beating and Machine houses. The two papermaking machines were overhauled and renewed with the addition of new strainers and pumps and the Board Machine was given a variable speed electric motor for the main drive.

On the converting side at Mugiemoss there was also considerable investment. This included a four colour aniline printer, a waxing machine for bread wrappers, and high speed Beasley French bag machines. Box making operations were expanded and a complete new factory finished in 1948 to manufacture railway containers.

A decision was also made to run a fleet of lorries in order to improve the firm's existing transport arrangements. By April 1948 three 8 ton lorries were in operation and others were envisaged.[6]

Major alterations to the Papermaking and Board Machines to increase their productive capacity substantially were added to the development programme rather later and the planned alterations were not carried out until the financial year 1950–51. The completion of the new power plant in 1948 was an essential pre-requisite, but the delay in increasing machine capacity also reflected the inadequate supplies of papermaking materials available to United Kingdom producers in the immediate post war years.

Waste paper collection, no longer stimulated by patriotic fervour and energetic local and central government support, fell to a much lower level, hence supplies available for consumption by the paper industry amounted to some 620,000 tons in 1946 and under 650,000 tons the following year, as compared with more than 750,000 tons each year from 1940 to 1942 with a peak of 867,000 tons in 1941. Even in 1948 and 1949 waste paper consumption by the industry was less than 800,000 tons. The Northern Wastepaper Company made strenuous efforts to boost Davidson's supplies, organising an efficient system of weekly collections from firms and individuals, but Frank Williamson on more than one occasion complained of shortages and denounced official apathy over promoting waste paper collection.[7]

The use of imported raw materials was affected by the acute dollar shortage. The machinery of Paper Control remained in being and until 1950 the Controller continued to import wood pulp, esparto, and pulp wood and to allocate supplies to individual firms on the basis of pre-war production figures. Imports of wood pulp, the main constituent, by 1946 were more than double the wartime nadir of 400,000 tons in 1944, yet in the three years 1946–48 the annual import figure was still scarcely more than half the pre-war total of over 1,600,000 tons, rising to nearly 1,200,000 tons in 1949.[8]

Until 1950 as can be seen from Tables 21 and 22 the output of paper and board at Mugiemoss was therefore relatively static. Production of paper in 1946 was some 24 per cent greater than in 1944, reaching a total of 5057 tons, but this figure was barely exceeded in each of the next three years, with a maximum of 5037

Table 21: Output of Paper, Machines Nos.2 and 3, 1945–1952

Year	Quarterly Output (tons)				Annual Output (tons)
	I	II	III	IV	
1945	1045	1096	975	1145	4261
1946	1164	1291	1206	1396	5057
1947	1267	1330	1139	1335	5071
1948	1431	1356	1297	1223	5307
1949	1240	1381	1134	1356	5111
1950	1365	1449	1490	1787	6091
1951	2130	2142	1883	1794	7949
1952	2247	1504	758	1513	6022

Source: Daily Report Books, 1945–52

Table 22: Output of Mill Board, Machine No.4, 1945–1952

	Quarterly Output (tons)				Annual Output (tons)
Year	I	II	III	IV	
1945	2674	3222	2800	3344	12,040
1946	3637	4019	3956	4453	16,065
1947	3365	4110	4137	4068	15,680
1948	4182	4239	4155	3993	16,569
1949	3836	3427	3647	4532	15,442
1950	5242	5464	5234	5961	21,901
1951	4189	5431	6761	6731	23,112
1952	7339	5417	2525	4669	19,950

Source: Daily Report Books, 1945–52

tons in 1948. The 1948 total moreover was precisely 50 tons greater than the pre-war peak in 1937.

The production of millboard, which had overtaken paper production in importance before the war, was clearly given preferential treatment in the immediate post war years. The data summarised in Table 22 shows that output in 1946 was slightly in excess of 16,000 tons. The figure was approximately a third higher than in 1944 or 1945 and easily surpassed the peak pre-war rate of production in 1939, which was equivalent to nearly 11,000 tons a year. However over the next three years 1947–49 there was no overall increase in output. In 1947 and 1949 the tonnage produced was slightly lower than in 1946 whilst in 1948 there was an increase of some 500 tons.

The alterations to the Paper and Board making machines planned for 1950–51 aimed to increase the output of paper by more than 50 per cent and millboard by at least the same extent. In July 1950 five drying cylinders and a suction press were added to No.2 machine with the intention of raising output to 66 tons in a normal working week, an increase of 50 per cent. In September it was the turn of No.3 machine which also received an additional five drying cylinders and a suction press as well as a new flow box and deckle aprons. This was intended to increase output by more than 66 per cent to some 132 tons per normal working week.[9]

The contract for the alterations to the Board Machine costing nearly £60,000 was placed with Walmsleys (Bury) Ltd in 1949. The firm supplied 39 drying cylinders with the necessary framing, gears and drive as well as a stack of calenders and also provided an erector to supervise the operation. The work began on 3rd March 1951 and the completed machine underwent its trial run at the end of April. The intention was to increase output to 528 tons per week soon after completion and to 660 tons by the end of 1951.[10]

Additions to the power plant at Mugiemoss planned for 1952 would increase the production of millboard still further. Thus the installation of an additional Turbo Alternator, providing increased steam pressure, it was estimated, would increase output to 792 tons a week.[11]

Unfortunately the increase in the capacity of the Paper and Board Making Machines coincided with a renewed shortage of papermaking materials. The devaluation of sterling by 30 per cent in September 1949, which was followed by a similar devaluation of Scandinavian currencies against the dollar, encouraged United States paper manufacturers to purchase Scandinavian wood pulp in much greater quantities. At the same time the expansion of the American economy, gathering force since the Autumn of 1949, generated a high level of demand for imported raw materials, even before the outbreak of the Korean war in May 1950 led to the deliberate stock piling of materials.[12]

Production of paper exceeded 1450 tons per quarter for the first time in the third quarter of 1950, rising to 1787 tons in the final quarter of the year. In 1951 the increased rate of production averaged nearly 2,000 tons per quarter, giving a total for the year of 7,949 tons, which represented an increase of nearly 56 per cent on the figure for 1949.

In the months preceding the major alterations to the Board Machine in the Spring of 1951 it had been possible to increase output of millboard to a much higher level. Output exceeded 5200 tons in each quarter of 1950, giving a figure of 21,901 tons for the year as a whole, which was an improvement of 32 per cent on the record output of 1948. In the third and fourth quarters of 1951, after the major alterations, output exceeded 6,730 tons per quarter – an increase of a further 23 per cent – rising in the first quarter of 1952 to a new peak of more than 7,300 tons, which in turn was 33 per cent higher than the average quarterly production in 1950.

In the United Kingdom and in other countries demand also increased markedly and as paper manufacturers competed for the limited supplies available the price of wood pulp soared, increasing three fold between the start of the boom and the middle of 1951. British mills were unable to run their machines to full capacity and there were growing delays in fulfilling orders from customers at home and overseas.[13]

Meanwhile, a surplus of waste paper in the first half of 1949, which had persuaded the Board of Trade in July to revoke the order compelling Local Authorities with a population of over 10,000 to collect waste paper, was followed by a serious shortage. Stocks were depleted as many Local Authorities ceased collection and it was not until November 1950 when the House of Commons debated the state of the paper industry that the Board of Trade with the aid of the Waste Paper Recovery Association launched a new salvage drive to encourage Local Authority action.[14]

In the circumstances it is hardly surprising that the additional output of paper and millboard at Mugiemoss was somewhat less than had been anticipated. The Chairman's statement to shareholders in the Annual Report in 1951 indeed declared that 'The real limit to our production and trade is the supply of raw materials'. Nevertheless, as can be seen from Tables 21 and 22, the increase in output was substantial.[15]

FORMATION OF THE DAVIDSON GROUP AND OTHER ADVENTURES

In the early post war years whilst internal expansion was restrained by the relatively static output of the Paper and Board Mills, a conscious decision was made to expand externally by acquiring firms which converted millboard into various types of boxes and containers. By entering this rapidly expanding field the Davidson directors sought to ensure a more secure and more profitable outlet for the production of their Board Mill, just as their predecessors had done for their Paper Mill in the later 19th century by embarking upon the manufacture and printing of paper bags.

In the space of less than three years three firms were taken over, all relatively cheaply. The first acquisition, in October 1945, was H.G. Smith (Boxmakers) Ltd of Peckham, London, with an issued capital of only £1,000. In March 1947 a controlling interest was secured in Charta Union Mill Ltd of Radcliffe, Lancashire, a firm which manufactured paper tubes, canisters, waxed paper, toilet rolls and folding and rigid boxes. Some £5,000 of the issued capital of £27,606 initially remained in the hands of outside investors and this was purchased piecemeal later as the opportunity arose.[16]

Fibreboard Boxes Ltd of Gateshead who made both rigid and folding boxes in a new factory erected on the Team Valley Estate in 1946 was acquired by an exchange of shares in September 1948. The entire issued share capital of £20,000 was exchanged for 120,000 five shilling ordinary shares in C. Davidson & Sons with a nominal value of £30,000. In addition in the preceding negotiations with the Fibreboard Boxes directors it was agreed as a further inducement to the deal that Fibreboard Boxes could declare a dividend of 20 per cent less tax for the year ended 31 March and pay the directors fees of £500.[17]

C. Davidson & Sons thus became the parent firm at the head of a group of companies producing paper and millboard and converting it into a variety of products. The other two subsidiaries in the group had older links with the parent firm. The sales organisation, as was noted in chapter 3, had been formed as a separate company in 1931, whilst the Northern Waste Paper Company had been in close co-operation since 1939 through the share holding of two directors of Davidsons and was wholly acquired in May 1946.[18]

Taken as a whole the group achieved some measure of vertical integration. The principal raw material, waste paper, was purchased and collected by the Northern Waste Paper Company, it was manufactured into paper and millboard at Mugiemoss, and these were converted there or in the three English factories into products required for wrapping and packing goods. Davidsons Paper Sales then sold the products to corporate customers who were often the final consumers.

The group might have been still larger had a less disciplined team of directors been in charge or had a masterful adventurer in the mould of Sir Eric Bowater been at the helm.[19] Three other possibilities for external expansion in 1948 and 1949 were considered and rejected either for specific reasons or simply because they did not fit into the broad general strategy and would overstretch the firm's financial resources.

In June 1948 the managing director and James Mearns Dawson examined a site at Cordale, Dumbartonshire, which was considered excellent for the erection of a Board Mill, but it was decided that no immediate action should be taken, the whole matter being 'one for long term planning'. Later in the year the Board had no hesitation in rejecting an offer to buy Ellangowan Paper Mills Ltd on the outskirts of Glasgow as a going concern. Davidson's managing director and the company secretary had visited the mills and they reported that:

1 The machinery was out of date.
2 The site was not good for further development if additional papermaking machines were required.
3 The finances of the company were 'extremely complicated and involved'.[20]

A proposal to erect a Board Mill in Eire, however, proved a more tempting project, involving lengthy negotiations before the final rejection. In December 1948 the managing director was approached to see if Davidsons would support the project. There was no other board mill in the Republic and it was indicated that the Irish government would not permit the erection of another board mill if the project went ahead, that tariffs would be used to protect the mill against cheaper imports, and that there would be no problem in finding the capital required. Davidsons investment in, say, a £600,000 company could be as low as £10,000 yet the firm would be given a controlling interest.[21]

The drawbacks to this attractive scheme, however, as the chairman, J.C. Duffus, pointed out were essentially two fold. If the firm invested only a small sum there would be relatively little income in dividends and it would be some 4 to 5 years before even this was received. On the other hand their own development programme needed the full attention of the management and would certainly suffer if the managing director had to make frequent visits to Eire to supervise the project. In addition J.M. Dawson was concerned that future political problems between Eire and the United Kingdom might lead to surrender of control over the new mill.

The managing director was confident that he could cope with the extra demands on his time by appointing a production manager for the Paper and Board mills at Mugiemoss to deal with day to day problems. At the same time the Irish negotiators were willing to consider granting an option to buy a block of shares in five years time, which would greatly improve the financial rewards.

At the final debate in October 1949 the Board was split on the issue. Dawson and Mellis supported the chairman, stressing the need to consolidate and concentrate upon the firm's own current development programme. The managing director, the secretary, and both Davidsons were in favour of the scheme, provided the share option was granted and the managerial problems were overcome. However, although there was a majority in favour it was felt that 'on an issue of this importance complete unanimity of the Board is essential' and the Irish government were therefore informed that the firm were unable to support the project.[22]

The following year there was a close encounter of a different kind. Eric Green, managing director of Hygrade Corrugated Cases Ltd, a Canadian firm's English subsidiary based at Southall Trading Estate in London, had approached Dawson to buy his Davidson shares, which would have given Hygrade a controlling interest. Dawson refused and informed his fellow directors, but over the next four months until the end of May discussions took place to explore the possibility of closer relations between the firms.[23]

Hygrade wished to ensure a secure British supply of kraft and test jute millboard in place of imports from Canada. The amount involved was considerable since Hygrade's output was estimated at some 23,000 tons and it was anticipated that this would increase when the latest corrugating machine was installed to perhaps 42,000 tons a year.

The provision of such a tonnage would involve Davidsons in substantial investment in addition to their existing commitments. A twin flow wire board machine 124 inches wide with auxiliary equipment, which was proposed at one stage of the discussions, for example, would have cost some £450,000. Hence the directors wanted to obtain a major injection of capital, short of surrendering control, sufficient to pay for all the extra plant needed and to contribute towards their current programme of development. An alternative proposal by Hygrade of a Consumers' Agreement, whereby they and some other users of these types of millboard would guarantee to take their future supplies from Davidsons, was not acceptable and after a visit to Mugiemoss by Green and a Canadian colleague early in May the negotiations came to an end.[24]

Ironically one of the most promising potential outlets for Mugiemoss paper developed by the Davidson directors involved investment in a firm which was not a member of the Group and in which the bulk of the capital was held by outside parties. The negotiations which led to the formation of Abertay Paper Sacks Ltd in December 1950 had begun in September 1948. H.V. Bonar of Low & Bonar the Dundee jute manufacturers, who were anxious to enter the expanding trade in paper sacks, met Frank Williamson and Peter Davidson to discuss the supply of Kraft sack paper for a machine which Low & Bonar were planning to install. By February 1949 Low & Bonar had been persuaded to drop their own plans in favour of a joint venture and negotiations were proceeding smoothly when it was learned in May that Jute Industries Ltd of Dundee also intended to commence paper sack production.[25]

This would have made life more difficult for the joint venture. If Davidsons refused to supply sack kraft paper to Jute Industries for their new plant Davidsons' other sales to the firm, which was a good customer, would be at risk, whilst if Jute Industries turned to other paper manufacturers for sack paper this would create the very competition that the Davidson Board had hoped to minimise. It was decided therefore that Jute Industries must be persuaded to participate in the joint venture and when Low & Bonar objected the Davidson directors disclosed the project to Jute Industries and finally overcame Low & Bonar's resistance to a tri-partite agreement by threatening to withdraw from the separate proposals already being considered with Low & Bonar.[26]

In August the three parties eventually agreed to collaborate in forming a separate company to manufacture paper sacks, each firm providing one third of the capital. Davidsons were to supply the Kraft paper and the other firms were to provide the sales organisation, but both selling and manufacture were to be provided as cheaply as possible so that the new company earned the major profit from the venture. Later that year it was decided that the capital initially should be limited to £50,000 – the maximum capital issue permitted without applying to the Capital Issues Committee – and that the new plant should be installed at Mugiemoss if a licence could be obtained for the erection of a factory to house it. A potentially serious source of future conflict was also removed by an agreement that if the company at some time wished to manufacture other products as well as the primary output of multi-wall paper sacks, the Davidson directors on the Board must give their approval if these products were already being made at Mugiemoss.[27]

In January 1950 the Dundee firms began to press for an early start to production. The Indian Government had stopped the export of *gunnies* which would create a shortage of jute sacks in the United Kingdom and enlarge the market for paper sacks. It was decided therefore to install the new plant in Dundee rather than at Mugiemoss, in premises owned by Jute Industries. Import licences for a Block Bottoming Machine from Germany and the Sewing Heads from the USA were obtained in March and in late April the layout of machinery and minor additions to the Dundee factory were approved, the lease being £550 a year.[28]

However it was December 1950 before the company, Abertay Paper Sacks Ltd, was finally launched with a capital of £50,000 in £1 shares. The name was the second preference because the Board of Trade objected to the name Scottish Paper Sacks Ltd as E.S. and A. Robinson of Bristol already owned a subsidiary company known as Paper Sacks Ltd. The six directors were drawn equally from the three founding firms, Peter Davidson and Duffus representing Davidsons' interest, an employee of Low & Bonar who had been sent to Canada for training was made foreman, and Jute Industries Ltd provided the company secretary.[29]

THE MEANS AND THE REWARDS

The collapse of the Hygrade talks in May 1950 emphasised the heavy demands that Davidsons development programme had made and was continuing to make upon the firm's financial resources. Up to the end of July 1950 modernisation and expansion had cost more than £450,000 – exclusive of regular maintenance and repairs – whilst it was estimated shortly after this date that additional expenditure of £180,000 had been or would be incurred before the current programme was completed.[30]

Part of the expansion was financed internally, encouraged by a taxation system which discriminated against distributed profits.[31] Yet extra capital was also raised on three occasions from 1945 to 1951.

In November 1945 the issued capital was doubled in size by the creation of 100,000 5 per cent cumulative preference shares of £1 each. These were offered at par to holders of the ordinary shares and the capital was soon subscribed.[32]

By May 1948 it was clear that additional outside capital would be needed. The company secretary presented a Budget Plan surveying likely future needs up to 31 January 1950 and it was decided to make an immediate application to the Capital Issues Committee, which advised the Treasury, for permission to issue £50,000 of ordinary shares in 5/- units at 12/6 each, thus raising £125,000, and to make a further application to the Committee should negotiations with Fibreboard Boxes Ltd reach a successful conclusion.[33]

Permission to issue 320,000 ordinary shares of 5 shillings was granted in August – 200,000 for cash at 12/6 each and 120,000 to be credited as fully paid and issued to the shareholders of Fibreboard Boxes Ltd in exchange for their entire issued share capital of £20,000.[34]

An extra-ordinary General Meeting in September 1948 approved an increase in the nominal capital of the company to £300,000 by creating 400,000 ordinary shares of 5 shillings, to be designated B shares to distinguish them from the existing 400,000 ordinary shares, which had been created by consolidating the 2,000,000 ordinary shares of 1 shilling in 1946 and which were now designated A shares. The B shares were to be converted on issue into B stock with voting rights of one vote per £1 of stock held. Of the B shares authorised 320,000 were to be issued immediately on the terms approved by the Capital Issues Committee and the balance of 80,000 issued at some future date.

The 200,000 B shares issued for cash were offered to the existing holders of A ordinary shares on the basis of one share for every two A shares held. By the end of October acceptances had been received for more than 197,000 of the shares and the remainder were readily disposed of on a pro rata basis to A shareholders who wished to purchase more than their allocation.[35]

Barely a year later the Board had begun to consider ways of raising additional capital. As an interim measure the North of Scotland Bank agreed to grant overdraft facilities of £100,000 for 6 months at 4 per cent.[36]

The original proposal in November and December 1949 was to issue £50,000 of Short Term Notes maturing in 5 – 10 years. The Glasgow Industrial Finance Corporation was willing to place an issue of £50,000 of four and a half per cent Notes at par for a commission of £300-£500. Alternatively the Corporation would consider underwriting an issue of £50,000 or even £100,000 of preference shares, the Board's second option, for a more substantial remuneration, and it was felt that this was a better choice since it would be difficult to raise extra capital in future if a series of Short Term Notes were outstanding.[37]

Matters were then suspended in view of the company's hopes of securing a large injection of capital through the Hygrade negotiations. Meanwhile the Bank increased overdraft facilities until 31 August 1950 to £150,000, which included the opening of a separate account through which purchases of pulp and waste paper would be paid.[38]

When discussions were resumed at the end of May 1950 it was felt that £200,000 should be raised at once, rather than £50,000 initially with more to follow. Quite apart from the demands of the current expansion programme the need for extra capital had been increased by the steep rise in wood pulp prices since the autumn of 1949, which meant that the bank overdraft limit of £100,000 would soon be very inadequate. The Board now favoured an issue of preference shares, if necessary at 6 per cent, but when their financial advisers suggested that a successful flotation might depend upon upgrading the return on the firm's existing 5 per cent preference shares, which could involve six months delay in securing approval from the Court of Session, it was decided to issue debentures instead.[39]

In November £300,000 of four and a half per cent mortgage debentures were issued at par. Half of them were offered to the general public and the other half were deposited with the Clydesdale and North of Scotland Bank as security for advances. The debentures were redeemable on 31 July 1970 or at the company's option from 31 July 1965 and were secured on the firm's heritable property at Mugiemoss, Aberdeen, and Glasgow. The title deeds of the Newcastle property were excluded in order to avoid the extra expense of employing English solicitors.[40]

The fruits of the high level of investment were seen in the substantial profits earned by the company in the post war years, which both pleased existing and potential future investors and generated some of the capital needed for further expansion. The data is summarised in Tables 23–25. In the five years 1946–50 the net income of the Davidson Group, consisting almost entirely of trading profits, ranged from nearly £126,000 in 1946 to some £207,000 in 1950, with an earlier peak of £198,000 in 1948. After setting aside a total of more than £164,000 for depreciation and providing for taxation, it was still possible to allocate nearly £100,000 to the General

Reserve Fund, contribute £3,500 to the Employees' Welfare Fund and set aside £13,000 in two other reserve funds.[41]

The shareholders saw no reason to complain. The 5 per cent dividend on the cumulative preference shares was regularly paid whilst the dividend on the ordinary shares rose from 15 per cent in 1946 to seventeen and a half per cent in 1947 and then remained at that level for the next three years.

When all these needs had been met there was still a healthy balance carried forward in the accounts to the following year. This ranged from less than £19,000 in 1946 to £85,000 in 1950, averaging £54,245 a year.

The subsidiary companies acquired since the war contributed in no small measure to the prosperity enjoyed by Davidsons' directors and shareholders. The separate data is not known for the financial year 1947–48 as the accounts at Companies Registration Office in Edinburgh are incomplete but it is clear that in the other four years from 1946 to 1950 the subsidiary companies as a whole accounted for nearly 46 per cent of the trading profits of the Group and in 1949 their contribution actually exceeded that of the parent company.

The full impact of the boom and inflation of 1950–51 was seen in the accounts for the year ending 28 July 1951, summarised in Tables 23–25. The turnover of the Group almost reached £2.5 million and profits soared to the unheard of heights of over £600,000. After provision of some £283,000 for taxation and £70,000 in depreciation, it was still possible to place £85,000 in the General Reserve and another £80,000 in the Stock Reserve and carry forward more than £103,000 to the next year's account. Meanwhile the ordinary shareholders saw the dividend level increased to 25 per cent.[42]

The chairman's statement issued in advance of the A.G.M. in December 1951 was suitably cautious, observing that 'The balance of the increased profit arises from favourable raw material purchases and other fortuitous circumstances which cannot be expected to recur'. Nevertheless he and his fellow directors must have felt fully justified in drawing up and executing the post war expansion plans which had created a prosperous and expanding group of companies capable of earning a volume of sales and a level of profits beyond the achievement of C. Davidson & Sons on its own.

Table 23: Net Income of the Davidson Group, 1946–1952

Financial Year Ending[1]	Trading Profits & Interest Received[2] £	Depreciation £	Provision for Taxation £	Debenture Interest £	Net Income Remaining £	Balance Forward from previous year £
1946	125,896	19,353	76,186	—	30,357	15,169
1947	163,709	19,571	82,264	—	61,874	18,708
1948	198,087	27,190	90,643	—	80,254	44,352
1949	156,421	41,375	57,927	—	57,119	54,597
1950	206,782	56,915	63,010	—	86,857	68,566
1951	620,737	70,141	283,802	4,769	262,025	85,004
1952	422,573	79,947	212,101	6,750	123,775	103,511

Source: *Annual Report and Accounts*, 1946–1952; *Printed Circular for Mortgage Debenture Issue*, November 1950

Notes: 1 In 1946 the financial year ended on Wednesday 31 July, in subsequent years on the Saturday nearest to that date, either at the end of July or at the beginning of August.

2 Includes profit on the sale of plant and after providing for administration (including directors' emoluments) and other expenses. Interest receipts exceeded £700 a year only in 1946.

Table 24: Appropriation of Net Income of the Davidson Group, 1946–1952

Financial Year Ending	Reserves: General Reserve £	Reserves: Others[1] £	Employees Welfare Fund £	Taxation Equalis'n Fund £	Dividends Less Income Tax £	Other Items[3] £	Balance Forward to next year £
1946	14,000	5,000	—	—	9,625	—	18,708
1947	13,500	2,000	1,500	—	12,375	8,429	44,352
1948	42,000	6,000	1,500	—	12,375	8,134	54,597
1949	10,000	—	500	—	20,075	12,500	68,566
1950	20,000	—	—	26,325	20,075	4,019	85,004
1951	85,000	80,000	2,500	20,296	26,312	15,992	103,511
1952	130,000	4,000	2,500	5,724	36,750[2]	—	116,085

Source: *Annual Report and Accounts*, 1946–1952; *Printed Circular for Mortgage Debenture Issue*, November 1950

Notes: 1 Preference Stock Dividend Reserve plus £2,000 Stock Reserve in 1947, £5,000 Pensions Reserve in 1948 and £80,000 Stock Reserve in 1951.

2 After the end of the financial year £100,000 of the Capital Reserves was capitalised and issued credited as fully paid in the form of £50,000 'A' Ordinary Stock and £50,000 'B' Ordinary Stock. This new stock qualified for the final dividend of 15 per cent approved by the AGM in December 1952.

3 Includes additional depreciation (1947–49), Goodwill written off (1950–51) and Debenture Stock issue expenses (£5,992 in 1951).

Table 25: Share Capital and Dividends, 1946–1952

Financial Year Ending	Ordinary Capital		5% Cumulative Preference Stock	
	Issued £	Dividend Per Cent	Issued £	Dividend Per Cent
1946	100,000	15.0	100,000	5.0
1947	100,000	17.5	100,000	5.0
1948	100,000	17.5	100,000	5.0
1949	180,000	17.5	100,000	5.0
1950	180,000	17.5	100,000	5.0
1951	180,000	25.0	100,000	5.0
1952	200,000[1]	25.0[2]	100,000	5.0

Source: *Annual Report and Accounts*, 1946–1952

Notes: 1 After the end of the financial year £100,000 of the Capital Reserves was capitalised and issued credited as fully paid in the form of £50,000 'A' Ordinary Stock and £50,000 'B' Ordinary Stock. This new stock qualified for the final dividend of 15 per cent approved by the Annual General Meeting in December 1952.

2 Excluding the operation referred to in Note (1).

THE END OF AN ERA AND A NEW BEGINNING

In August 1951 Colonel Thomas Davidson died after a short illness at the age of 71, having served continuously as a director since 1911. His death broke the last link with an era when the family firm was still the ubiquitous and dominant form of business organisation, even though larger and more powerful competitors had begun their predations. When he joined the Board virtually all the directors bore the Davidson name, when he died his only son Peter was the sole remaining member of the family serving as a director. He must have been proud that his son was the chief executive but very conscious that Peter was 41, unmarried and likely to remain a batchelor, thus ending the Davidson connection.[43]

The traditional family firm when faced with a dearth of sons to carry on the business had often promoted to the Board established staff who had given loyal service. However the three additional directors appointed a few days before Colonel Davidson's death had virtually no previous experience with the company.

Two were local businessmen. B.W. Tawse was the managing director of William Tawse Ltd, Public Works Contractors of

Aberdeen, and a director of nine other firms. H.R. Spence, Conservative M.P. for West Aberdeenshire since 1945, was chairman of Harrott & Co Ltd, an Aberdeen hosiery manufacturer, and the managing partner in William Spence & Son a knitwear manufacturer of Huntly, Aberdeenshire.[44]

The third appointee was Eric J. Warburton who had been recruited from a Canadian firm as mill general manager in March 1950. He had considerable experience of the paper industry, including some years at St. Anne's Board Mill, Bristol, one of the leading mills in that sector of the industry.[45]

Appropriately, the chairman, J.C. Duffus, was now the longest serving member of the Board. He was the last of the triumvirate who had rescued the firm from near extinction in 1935.

Frank Williamson had resigned as managing director and left the Board in January 1950 to take charge of the new project in Eire, subsequently becoming the first managing director of National Board and Paper Mills Ltd, the company floated to finance it. His departure reflected strained relations with Duffus over that project and earlier over the making of staff appointments and his resignation undoubtedly strengthened the position of Duffus on the Board.[46]

The fourth new appointment in 1951 probably strengthened the chairman's hand still further. R.M. Ledingham, who was co-opted in December, was an Aberdeen advocate with an office in Golden Square near to Duffus' own law firm. Both men were almost the same age, Ledingham, who at 58 was the second oldest director, being 18 months younger than his chairman.[47]

The new members who joined the Board in August 1951 found themselves at the beginning of a collaborative venture with one of the firm's largest customers which was to have far reaching consequences. One of the leading products of the Board Mill was plasterboard liners. In the seven years 1946 – 1952, as can be seen from Table 26 this never constituted less than 21 per cent of its output and in two of the years, 1947 and 1948, almost half of the mill's output was in this form. British Plaster Board Ltd had been a customer since the infancy of the Board Mill and in post war years sales to the firm became of increasing importance as British Plaster Board came to dominate the United Kingdom market, accounting by 1951 for some 70–75 per cent of U.K. production of plasterboard.[48]

Table 26: Output of Plasterboard Liners, 1946–1952

Year	Output		Percentage of Board Mill Output
	Tons	Index Nos.[1]	
1946	6,417	100	40
1947	7,525	117	48
1948	8,003	125	48
1949	3,213	50	21
1950	6,606	103	30
1951	6,764	105	29
1952	5,534	86	28

Source: Daily Report Books, 1946–52
Notes: 1 1946 = 100

The acquisition of Gyproc Ltd by British Plaster Board in 1944 gave that firm interests in S. Africa. Subsequent expansion in the South African market led quite naturally to thoughts about the possibility of beginning production further north in Southern Rhodesia where accelerating economic growth seemed to offer very favourable opportunities. Action was precipitated in 1951 when it was learnt that a company had been formed by Hans Leobbecke to erect a paper mill with a single papermaking machine at Umtali (now known as Mutare) in Southern Rhodesia. The Premier Paper Company of S. Africa had made a takeover offer but Leobbecke was willing to enter into discussions with British Plaster Board's South African subsidiary and by their invitation the Davidson Board as well.

J.C. Duffus was keen to participate provided closer examination proved the project was viable, even offering to find the cash Leobbecke needed to pay a bill of £2,500 for materials and machinery, due in August 1951, if Leobbecke agreed to talk to British Plaster Board and Davidsons before reaching any decision on the Premier Paper Company's approach. Duffus believed that the S. African company had connections with one of the larger paper groupings in the United Kingdom, who would oppose Davidsons' participation, and he pointed out to his fellow directors that British Plaster Board might decide not to contest the Premier Paper Company's bid if it was permitted either to purchase part of the equity or if it was guaranteed sufficient supplies of plasterboard liner to meet its requirements.[49]

After visits to S. Rhodesia in September and October 1951 it was decided to make a definite commitment. A report by Duffus and R. Jukes, a director of British Plaster Board, which was submitted to both Boards of directors concluded that the long term prospects for a mill in S. Rhodesia were good and that a single Board Machine should be installed initially, capable of producing the tonnage which British Plaster Board required with a small surplus to be disposed of to other customers. Soon after this it was agreed that a holding company should be formed with a capital of £270,000 to acquire the whole of the issued share capital of Umtali Paper Mills Ltd and to operate a plasterboard plant using its liners at Bulawayo. C. Davidson & Sons were to provide £40,000 of this sum, British Plaster Board and its S. African subsidiary £100,000 with the balance being found locally. Leobbecke was appointed general manager of the Board Mill and negotiations proceeded to secure local timber supplies for the production of wood pulp whilst arrangements were made to mine gypsum deposits discovered in the south of Northern Rhodesia to supply the plasterboard plant.[50]

The discussions leading up to the joint venture strengthened the belief among the negotiators that closer long term co-operation between C. Davidson & Sons and British Plaster Board was desirable. In the official internal history of BPB published in 1973 it is recorded that the initial move towards an amalgamation was taken by R.S. Jukes on a return journey from South Africa with J.C. Duffus when their plane was delayed with engine trouble at Kano in Nigeria in September 1952.[51]

Jukes recalled that:

> I had been waiting for the opportunity to put it to J.C. that we should get together on a more solid basis. As we walked around the airfield – it was one of those marvellous clear African nights – I simply said 'what about amalgamation between our two companies?'

However it is clear that a romantic atmosphere and a warm personal friendship were on their own insufficient to produce a permanent union and that solid mutual benefits and shrewd calculation lay behind the creation of the successful business marriage.

The talks about a merger were indeed protracted and when the final terms of British Plaster Board's offer to the shareholders of C. Davidson & Sons were revealed in April 1953 the *Aberdeen Press and Journal* commented that the first approach – a proposal to buy a stake in the firm – had been made about a year earlier, in other words before the stroll at Kano airport, and had been rejected by the Davidson Board.[52]

However a marked change in the economic climate facing the paper industry during the course of 1952 underlined the dangers of complete independence and almost certainly influenced attitudes towards joining a larger economic organisation. The chairman and his fellow directors were now facing a very different situation from that which had prevailed since 1939.

For years restrictions on the import of papermaking materials had held back the output of the paper industry below pre-war levels and demand had been artificially restricted, with customers having to accept delays and to limit the volume of their purchases. The boom of 1949–51, noted in section 1, which had pushed up wood pulp and paper prices to a record level, marked the end of this period. During 1952 a substantial fall in prices spread over several months led customers to hold back their orders in anticipation of further reductions in the price of paper products. The marked fall in demand for the first time since 1939 sent shock waves throughout the paper industry. An editorial in the *Papermaker* in December 1952 declared gloomily:

> December 31 1952 will surely register the close of one of the most depressing years ever experienced in modern times by the paper trade. Although the extra-ordinary demand for paper during the previous two years was a source of constant misgiving to all who had spent many years in paper making and paper selling, few anticipated that the ease in buying, when it arrived, would be so drastic as to almost present a complete cessation of purchasing.[53]

During the recession some mills closed for short periods and others went onto short time working, reviving memories of the 1920s and 1930s. Overall the output of paper declined by more than 20 per cent between 1951 and 1952, whilst the production of millboard declined by nearly 13 per cent. Less tangible but no less important was the message absorbed throughout the industry, that a more competitive environment had once more returned.[54]

At Mugiemoss the decline in output was slightly greater. The production of paper fell from 7,949 tons in 1951 to 6,022 tons the following year, a decline of more than 24 per cent. The production of millboard fell from 23,112 tons to 19,950 tons, a reduction of nearly 14 per cent. For both the paper and board mills the level of output was lower than in 1950, whilst in the third quarter of 1952 the output of paper sank to the lowest figure since 1938 and the production of millboard was lower than in any quarter since the spring of 1943.[55]

Viewing these figures J.C. Duffus must have felt that the caution he had expressed in May 1950, when the final stages of the development programme for the Board Machine were approved, was fully justified. He had agreed that the capacity of the machine should be increased to nearly 700 tons a week by the end of 1951 but voiced doubts about the ability to sell the additional output if further alterations proposed raised the capacity to 1,000 tons a week. If the machine had then to be operated below its optimum capacity this would increase costs per ton and would be uneconomic. However Peter Davidson had vigorously defended the bolder scheme and after some discussion this was adopted.[56]

The financial effects of the recession were seen in the next two sets of accounts presented to shareholders. The annual directors report and accounts for the financial year ending 2nd August 1952 reflected the impact of only the start of the depression. Group trading profits declined by nearly a third from the record figure of £620,737 in 1951 to £422,573 – mainly because the value of stocks of raw materials and finished goods was reduced substantially as prices fell – and after meeting depreciation, taxation, debenture interest and directors' emoluments this was reduced to less than £110,000. Utilising the Stock Reserve Account of £82,000, set up specifically to compensate for a sharp fall in the price of stocks, increased the sum available for distribution and £103,511 had been brought forward from the previous year. Hence after paying out nearly £37,000 in dividends, contributing £2,500 to the Employees' Welfare Fund, and increasing the Taxation Equalisation Fund and the Preference Stock Dividend Reserve by £5,724 and £4,000 respectively the directors were able to add £130,000 to the General Reserve Fund and still leave more than £116,000 to carry forward to the next financial year.[57]

The next set of accounts presented to shareholders, covering a period of eight months to 31 March 1953, showed a record balance of £154,314 to be carried forward to the coming financial year. However a closer examination reveals that this figure owed little to the current trading position.[58]

The accounts were credited with an additional £75,015 derived by taking £18,000 set aside for the Excess Profits Levy which was no longer required and by reducing the Reserve Fund for Income Tax for 1953–54 by £57,015. This was then used to write off £62,323 in Goodwill and £3,000 from the value of 2.5 per cent Treasury Stock, leaving a sum of £9,692.

The group trading profits before depreciation and taxation amounted to £129,629 and £1,439 was received in interest. After deducting £53,683 for depreciation, £7,631 for directors' emoluments, £4,500 for debenture interest, and £11,020 paid in profits tax the sum remaining amounted to £54,234. Dividend payments of £25,696 then reduced the amount to only £28,538. Hence the bulk of the sum carried forward to next year's accounts was derived from the substantial balance brought forward from the previous financial year.

Moreover the group trading profits before depreciation and taxation were equivalent to an annual total of some £194,000. This figure was lower than the level achieved in the financial years 1947–48 or 1949–50 before the planned increase in productive capacity at Mugiemoss had been completed.

Perhaps the most discouraging performance was recorded by Abertay Paper Sacks Ltd where Davidsons were in a minority position. Production finally began in August 1951, more than eight months after the company's formation, and the first set of accounts to 30 September 1951 showed a loss of £434. However this figure was an understatement since all overheads before the plant began operations were charged to 'Work in Progress'.[59]

The accounts for the next financial year, which reflected both a continuing shortage of Kraft paper for sack making and also the general recession in the paper industry, made even more dismal reading. A loss of £30,061 was recorded on trading operations and when allowance was made for depreciation of £4,469 and for a proportion of the expenses of company formation plus the previous year's loss, the cumulative loss carried forward in the balance sheet exceeded £35,000. The bank overdraft had risen correspondingly to £23,160 from £1,028 at 30 September 1951 whilst the share capital of £49,998 in issued shares had been supplemented by raising a loan of £36,000.[60]

The long term future potential of the company as an outlet for Davidsons' Kraft paper and as a source of profits was perhaps not in doubt. Nevertheless the return on a one third share in an investment of some £86,000 up to the end of 1952 was scarcely encouraging.

Overall as C. Davidson & Sons emerged from the depression no serious signs of financial strain were apparent. Contracts for future capital expenditure were at a record level of more than £198,000 in March 1953 but the Capital and Revenue Reserves amounting to some £640,000 were more than sufficient to cover these commitments, to meet most unforeseen liabilities which were likely to arise, and to cushion shareholders against a fall in dividend if there was a further recession in the near future.[61]

The confidence of shareholders in the directors was scarcely in doubt. The increased dividend of 25 per cent on the Ordinary Stock received in 1951 was maintained in 1952 and the dividend proposed for the eight months to 31 March 1953 – at the rate of 14.5 per cent – was equivalent to nearly 22 per cent in a full year.

Moreover the directors decided to capitalise £100,000 of the Capital Reserves by issuing £50,000 of 'A' Ordinary Stock and £50,000 of 'B' Ordinary Stock, which was credited as fully paid up, to existing stockholders in the ratio of one new unit for every two already held in these categories. The new shares were issued in September 1952 and participated in the dividend of 14.5 per cent paid on all ordinary stock for the eight monthly period ending on 31 March 1953.[62]

If there was no financial imperative propelling the firm towards merger with British Plaster Board, there were nevertheless some powerful reasons for it. BPB was one of the firm's largest customers and as we have seen more than a quarter of the output of the Board Mill consisted of plasterboard. The recession of 1952 whilst not inflicting major damage must have given the directors pause for reflection about the firm's future in a more competitive environment. Certainly it strengthened the chairman's earlier worries about the ability to sell the greatly increased output.

There was also a continuing need for substantial investment each year in order to remain competitive. Joining a larger organisation which was more familiar to investors in the United Kingdom and therefore able to raise external capital more easily and more cheaply, might make more rapid expansion possible.

The key personalities were also important. As the most powerful member of the Board the collaboration and friendship of J.C. Duffus with R.S. Jukes of British Plaster Board paved the way. The support of Peter Davidson, the chief executive, was crucial and as an established batchelor he had no dynastic ambitions to hold his own personal enthusiasm in check.

If any doubts remained on the Board or among local shareholders and employees these were allayed by the assurances given about the future of Mugiemoss. It was reported in the Aberdeen *Press and Journal* in April 1953 that if the takeover bid was successful British Plaster Board intended to carry on Mugiemoss Works under the present management and that they would do all they reasonably could to maintain it as a local industry giving local employment. Reflecting this attitude, when the takeover finally occured six of the eight Davidson directors, including Duffus and Peter Davidson, had seats on the new Board of Directors, together with R.S. Jukes, L. Biggs and W.S. Trimble, directors of British Plaster Board.[63]

The final terms offered to the Davidson shareholders were nicely judged. One 5 shilling unit of British Plasterboard Ordinary Stock was to be given for each 5 shilling Ordinary A and B unit of Davidsons, plus one 7 per cent Preference Share of £1 for every 40 units of Davidsons A and B Stock. Preference shareholders were offered three £1 7 per cent cumulative preference shares in BPB in exchange for four £1 5 per cent cumulative preference shares in their own firm.

In the weeks before the offer was made the BPB Ordinary Shares had been trading on the Stock Exchange at between 14s 1½d and 15s as compared with 12s 6d to 13s 6d for Davidsons Ordinary Stock, whilst the comparative figure for the preference shares was 26s as against 16s 3d. Hence although the deal was not very generous it did offer a positive financial gain, quite apart from the prospect that joining a larger organisation might boost future expansion and generate still higher profits.[64]

By 21 April 1953 the holders of 97.5 per cent of Davidsons' Ordinary Stock and of 87 per cent of the preference shares had followed the recommendation made by the directors and accepted the BPB offer. A month later the holders of BPB stock voted to approve the acquisition and the independent existence of C. Davidson & Sons came to an end – 78 years after the firm had been first floated as a limited company and 142 years after Charles Davidson, the founder, had begun operations at Mugiemoss.[65]

Chapter 6

The Shareholders

The continued expansion of C. Davidson & Sons in the final quarter of the 19th century, as we have seen, was dependent partly on the ability of the firm to attract additional capital from the general public. The Davidson brothers might well have kept the firm entirely within their own hands, tapping alternative sources of finance as many industrial firms continued to do, but their firm would probably have grown more slowly, whilst in the increasingly competitive years after 1900 the broader base of shareholder capital brought greater stability and security as the directors struggled to maintain the existing volume of business.

It is the aim of this chapter to examine the background and character of these shareholders. The size of their holdings, their occupations, and the areas in which they lived will be considered in turn.

The conversion of the family firm into a limited company in February 1875 was designed for the convenience and security of the Davidson brothers rather than for any urgent need to secure extra capital for expansion. The four brothers each received 975 fully paid up shares of £10 as part of the purchase price paid by the company for the existing business and by October 1877 the issued share capital of the company was still only £50,000 of which the brothers Alexander, John and David, who were the three directors, held £35,600–71 per cent of the total – whilst executors acting for George Davidson, the other brother, on behalf of his widow, held a further £2,000.[1]

Over the next few years as the share capital was increased to provide more capital for expansion the proportion of the equity held by the directors declined. By October 1883 when the issued capital, now divided into £1 shares, had increased to £60,000, the three directors had added some £8,900 to their holdings which now stood at 44,518 shares – 64 per cent of the total. However by October 1891 when 96,366 shares had been issued the number of shares held by the directors had fallen to 24,468 – a mere 25 per cent – and it is clear that the directors had used the opportunity provided by the substantial inflow of additional capital from the public both to sell shares and to transfer others to members of the family who were not in control of the firm. In 1891, for example, Alexander Davidson's three sons and his wife living together in Wimbledon each held 200 shares, whilst Charles William Davidson, the son of Charles Davidson – a former partner, who died in 1870 – was recorded with 112 shares.[2]

The combined share holding of 25 per cent in 1891 nevertheless was more than sufficient to retain control of the firm in the hands of the directors. The number of large share holdings, defined here as persons owning shares of the nominal value of over £300, exclusive of the directors increased from 11 in October 1877 to 19 six years later and to 30 in October 1891, whilst their combined share ownership over the 14 years increased from 14 per cent to 22 per cent of the total. However none of these individuals could remotely match in size the shareholding of a single director, even though John Davidson had reduced his stake in the firm to 4,847 shares by October 1891. In that year George Anderson and Charles A Mollyson, bankers of Aberdeen, had a joint holding of 5,256 shares but the next shareholdings in size, belonging to Thomas Fraser and James Ogston of Aberdeen, were both less than 1200 shares. Alexander Milne Ogston living near Aberdeen possessed 950 shares and three other persons held more than 800 – Robert Falconer a Stonehaven solicitor, John Hird of Buxton, and Gordon Pirie of Alexander Pirie & Sons Ltd, paper manufacturers of Stoneywood.

In earlier years the disparity between the directors' holdings and those of other large shareholders was still more marked. In October 1883 Charles Cook of Aberdeen held 1500 shares of £1 but the second largest holding, that of James Alexander Beattie, an Aberdeen land surveyor, was 780 shares. The executors of David Fiddes, an Aberdeen doctor, who held 770 shares, and Robert Ross Robertson of London who held 720 ranked as third and fourth in importance, whilst there were four holdings in the 600–650 range.

Six years earlier the largest shareholder without a seat on the

Board was in fact the widow of George Davidson, the fourth brother, whose holding was noted earlier. Charles Cook, then described as an hotelkeeper of Ballater, held 96 shares of £10 whilst the next individual shareholdings in size – both of 55 shares – belonged to James Saint, junior, a draper, and David Fiddes, a doctor, both of Aberdeen.

The majority of shareholders had invested much smaller sums, as can be seen from Table 27. Even in 1877 when the capital consisted of £10 shares 33 per cent of shareholders held shares with a nominal value of £50 or less. The conversion of the equity into £1 shares in February 1883, the issue of additional £1 shares subsequently, and the favourable impression created by the high level of dividends paid during the 1880s, encouraged investors of relatively moderate means to purchase the company's shares. In October 1883 38 per cent of the shareholders held shares with a nominal value of £50 or less and by October 1891 this proportion had increased to nearly 53 per cent.

The growing numerical importance of relatively small shareholders may be expressed in a slightly different way. Overall the number of shareholdings increased from 85 in 1877 to 195 in 1883 and 758 in 1891. The dividing line separating the bottom third of these from the wealthier two thirds of shareholders declined from over £50 in 1877 to under £30 fourteen years later, whilst the median value, dividing the shareholders into two halves, fell from more than £100 in 1877 to £70 in 1883 and less than £50 in 1891.

Investors in the lowest category, holding shares with a face value of no more than £20, also grew in importance. The number of such shareholders increased from 13 to 161 between 1877 and 1891 whilst proportionally they represented 21 per cent of shareholders in 1891 as compared with 15 per cent in 1877. However although their ranks included some persons of relatively humble social status, as we shall see, share ownership even on this modest scale was clearly beyond the aspirations and means of the vast majority of the population at the end of the 19th century, from the skilled workman earning £2 a week to the agricultural labourer struggling to survive on less than half that amount.[3]

Unfortunately the occupations of numerous shareholders were not recorded. It is tempting to assume that the absence of an occupational description in the case of a male shareholder indicated that

Table 27: Size of Shareholdings, 1877–1891

No. of Shares held of £1[1]	13 October 1877		10 October 1883		10 October 1891	
	Number of Shareholders	Per Cent	Number of Shareholders	Per Cent	Number of Shareholders	Per Cent
1–10	8	9.41	12	6.15	52	6.86
11–20	5	5.88	23	11.79	109	14.38
21–30	1	1.18	15	7.69	97	12.80
31–40	3	3.53	5	2.56	69	9.10
41–50	11	12.94	19	9.74	73	9.63
51–60	1	1.18	4	2.05	48	6.33
61–70	2	2.35	20	10.26	35	4.62
71–80	—	—	5	2.56	31	4.09
81–90	—	—	3	1.54	11	1.45
91–100	4	4.71	14	7.18	36	4.75
101–110	11	12.94	6	3.08	10	1.32
111–120	2	2.35	5	2.56	23	3.03
121–130	1	1.18	2	1.03	24	3.17
131–140	—	—	13	6.67	6	0.79
141–150	1	1.18	2	1.03	13	1.72
151–160	6	7.06	2	1.03	11	1.45
161–170	—	—	2	1.03	8	1.06
171–180	—	—	1	0.51	6	0.79
181–190	—	—	1	0.51	6	0.79
191–200	2	2.35	9	4.62	17	2.24
201–210	1	1.18	3	1.54	1	0.13
211–220	8	9.41	3	1.54	6	0.79
221–230	—	—	1	0.51	7	0.92
231–240	—	—	—	—	2	0.26
241–260	2	2.35	1	0.51	10	1.32
261–280	2	2.35	2	1.03	6	0.79
281–300	—	—	—	—	8	1.06
over 300	14	16.47	22	11.28	33	4.35
TOTAL	85	100.00	195	100.00	758	100.00

Source: CRO, Edinburgh, Davidson Files, Annual return of shareholders and capital, 1877, 1883, 1891.
Notes: 1 In 1877 the shares were in £10 denominations.

the person was of independent means and should therefore be assigned to a high social class. However this assumption may be too simple, for the extent of omissions increased markedly as the number of shareholders increased, perhaps indicating an economy of clerical effort by the recorder of the data rather than any real upward movement in social status. Thus in 1877, when the firm's equity consisted of fully paid up shares of £10, 13 per cent of male shareholders were recorded without an occupation, whereas in 1883 although the share denomination was only £1 the proportion had risen to nearly 35 per cent and in 1891 the proportion exceeded 54 per cent.

It is apparent from Table 28 that more than half of the male shareholders whose occupations are known were recorded in Commerce or the Professions. Merchants and retailers accounted for nearly 38 per cent of the total in 1877, declining to some 22 per cent in 1891 whilst the importance of the professions increased from under 23 to more than 38 per cent.

Amongst the professions the most striking changes were the increase in the proportion of medical men from 3 to 10 per cent and the very rapid increase in ministers of religion who accounted for over 9 per cent of the total by 1891. Correspondingly the relative importance of persons providing financial services such as banking, stock broking and accounting halved.

The range of professional services recorded also increased. By 1891 7 per cent of shareholders were professional men who operated outside the fields of law, medicine, finance, and religion and their ranks included an optician, an auctioneer, an architect, and a journalist.

Businessmen and farmers constituted a relatively small proportion of shareholders. Farmers never represented more than 7 per cent of the total and the two groups combined at their peak in 1877 and 1883 formed approximately 17 per cent, falling to 10 per cent by 1891. Paper manufacture not surprisingly was the most common type of industrial enterprise and in addition to the Davidson directors entrepreneurs from two other firms in the industry held shares – Gordon Pirie of Alexander Pirie and Sons Ltd, who was noted earlier, and Gilbert Johnson Wildridge, then general manager and soon afterwards a partner in Robert Craig & Sons, Moffat Mills, Airdrie.[4]

In 1877 skilled clerical and supervisory employees of all kinds from manager, foreman, clerk, cashier and commercial traveller to postmaster and excise officer, comprised the third largest group with 18 per cent of the total. However by 1891 although some additional occupations were recorded, including two army officers, a commander in the Royal Navy, and two policemen above the rank

Table 28: Occupations of Male Shareholders, 1877–1891

	Percentage of Shares Held		
	1877	1883	1891
1. *Commerce*			
Merchant	28.79	17.24	11.68
Retailer	9.09	12.65	10.75
	37.88	29.89	22.43
2. *Professions*			
Financial	15.15	8.05	7.01
Legal	3.03	2.30	4.67
Medical	3.03	4.60	10.28
Ministers of Religion	—	1.15	9.35
Others	1.52	4.60	7.01
	22.73	20.70	38.32
3. *Businessmen*			
Manufacturers	7.58	5.75	4.21
Ship Owners	3.03	2.30	0.93
Others	3.03	2.30	—
	13.64	10.35	5.14
4. *Agriculture*			
Farmer	3.03	6.90	5.14
Others	1.52	2.30	1.40
	4.55	9.20	6.54
5. *Skilled Workers: Clerical and Supervisory*			
Manager/foreman	4.55	3.45	1.40
Public Servant	4.55	1.15	2.34
Military/Naval Officer	—	—	1.40
Ship's Master	—	—	0.93
Clerk/Cashier	4.55	4.60	2.80
Commercial Traveller	4.55	1.15	0.93
Others	—	1.15	—
	18.18	11.49	9.81

Table 28 – *continued*

	Percentage of Shares Held		
	1877	1883	1891
6. *Skilled and Semi-Skilled Workers:*			
Manual			
Engineer/Millwright	1.52	2.30	3.27
Building Trades	1.52	5.75	3.27
Household and other Services	—	8.05	5.61
Others	—	2.30	5.61
	3.04	18.40	17.76

Source: See Table 27

of sergeant, the relative importance of the group had declined to 10 per cent.

Displacing the group in importance were men who may be described as skilled and semi-skilled manual workers. They accounted for only 3 per cent of shareholders in 1877 but some 18 per cent in 1883 and 1891. However it is impossible to determine how many of the group – for example men recorded as shoe maker, cabinet maker, upholsterer, or among the building trades as carpenter, slater, and painter – were self employed and some may have been retailing as well as making products. Hence the relative increase in their number does not necessarily indicate any extension of share ownership among the working class. Nevertheless in 1891 seven coachmen, three butlers, and a groom – described in the Table as providers of household and other services – were recorded, as compared with none in 1877, and together they represented nearly 6 per cent of male shareholders whose occupations are known.

Overall it is clear that the male shareholders were drawn mainly from the higher social classes. Excluding men recorded without an occupation, some of whom certainly were of independent means, merchants, retailers, businessmen, professional men, and farmers constituted not less than 77 per cent of the total in 1877 and some 68 to 70 per cent in 1883 and 1891. Skilled and semi skilled workers exclusive of agriculture, accounted for 21 per cent of shareholders in 1877 and for very nearly 30 per cent in 1883, but as we have seen some of them may have been self-employed. Unskilled workers were not recorded at all.

A substantial proportion of the shareholders were female. Excluding shareholdings recorded jointly in the name of both husband and wife the proportion increased from 13 per cent in 1877 to 25 per cent in 1883 and to 32 per cent in 1891. In the two latter years over half the female shareholders – 57 and 51 per cent respectively – were single, whilst widows accounted for 15 per cent in both years and married women for 28 and 33 per cent.

The growing proportion of shareholders who were female however was reflected only partly in their importance as a source of capital for the firm, since the average size of holding was smaller than that of male shareholders. Thus, excluding the three Davidson directors, in 1883 the average male share holding was 150 shares as compared with 79 whilst in 1891 the figures for male and female shareholders were 107 and 68 respectively. Viewed as a whole female shareholders held less than 6 per cent of the issued share capital not held by the directors in 1883, rising to 16 per cent in 1891.

Only a minute fraction of female shareholders were recorded with occupations. Even in 1891 a mere 8 women were noted in gainful employment – 3 teachers, 2 milliners, a dressmaker and two domestic servants – all of them single. In view of the limited employment opportunities for middle class women and the prevailing attitude that the woman's place was in the home caring for her family, these figures are scarcely surprising.[5]

Shareholders of both sexes were mainly local persons. It is evident from Table 29 that in 1877 and 1883 some 87–89 per cent resided in North East Scotland within a radius of 50 miles of the company's head office at Mugiemoss. Indeed 56 per cent or more of the total resided less than 11 miles away, the majority of these living inside the modern boundaries of the city of Aberdeen.

The issue of additional shares after 1883 attracted proportionally more shareholders from beyond North East Scotland. Nevertheless 76 per cent of investors in 1891 still resided within a 50 mile radius and 49 per cent lived within 11 miles of Mugiemoss, mostly in Aberdeen.

Beyond the immediate vicinity of Aberdeen shareholders were to be found throughout North East Scotland. In 1891, for example, shareholders were drawn from over 60 different places between 11

Table 29: Location of Shareholders' Residences, 1877–1891[1]

Place of Residence	October 1877 Number	October 1877 Per Cent	October 1883 Number	October 1883 Per Cent	October 1891 Number	October 1891 Per Cent
I *North East Scotland* Distance from Company HQ (miles)						
0–10	47	55.95	109	57.67	362	48.92
11–20	11	13.10	20	10.58	53	7.16
21–30	12	14.29	24	12.70	90	12.16
31–40	3	3.57	11	5.82	52	7.03
41–50	—	—	5	2.65	9	1.22
Total	73	86.90	169	89.42	566	76.49
II Northern Scotland	—	—	2	1.06	27	3.65
III Central & Southern Scotland	3	3.57	4	2.12	41	5.54
SCOTLAND TOTAL	76	90.47	175	92.59	634	85.68
IV *ENGLAND*						
1. Northern	—	—	2	1.06	17	2.30
2. North Wales	—	—	—	—	2	0.27
3. Midlands	2	2.38	2	1.06	10	1.35
4. London Area	6	7.14	6	3.17	56	7.57
5. South East England	—	—	—	—	10	1.35
6. South West England	—	—	1	0.53	6	0.81
ENGLAND TOTAL	8	9.52	11	5.82	101	13.65
V *OVERSEAS*	—	—	3	1.59	5	0.68
	84	99.99	189	100.0	740	100.01

Source: See Table 27
Notes: 1 Excludes shares held by a deceased person's executors. In the few cases where shares were held jointly, generally by husband and wife, only one person has been 'counted' at that address.

and 50 miles from Mugiemoss. The typical contribution made by small individual communities where the population numbered at most a few hundred persons ranged from one to three investors, whilst larger centres relatively near to Aberdeen, like Stonehaven and Laurencekirk, provided a dozen or more. However Turriff supplied 18 investors and the fishing community of Banff 40 miles from Mugiemoss no less than 30.

Other regions of Scotland including the populous and prosperous Central Region embracing Edinburgh and the Glasgow connurbation, where there were ample industrial and commercial outlets for investment, accounted for less that 4 per cent of shareholders in 1877 and 1883. However by 1891, perhaps reflecting a growing awareness by stockbrokers and other financial advisers of the firm's consistently high dividend payments, the contribution made by the rest of Scotland had increased to over 9 per cent.

English investors comprised under 10 per cent of shareholders in 1877 and 1883, rising to nearly 14 per cent in 1891. More than half of them were drawn from the London area and the adjacent south eastern counties, reflecting both the concentration of wealth and population in the Greater London area and also the strong presence maintained by Davidsons in the London market.

The Midlands accounted for just over 1 per cent of shareholders in 1883 and 1891, the South West of England's contribution was even smaller. Yet the North of England and North Wales were not much more important as a source of shareholders, contributing less than 3 per cent of the total in 1891 and far less earlier, although the activities of the branch offices in Newcastle on Tyne and Liverpool might have been expected to generate some interest among potential investors.

Investors living overseas accounted for a tiny fraction of shareholders – 1.59 per cent in 1883 and 0.68 per cent in 1891. Of the five individuals recorded in the latter year two were living in South Africa, one in Cuba and two, a merchant and a minister with a medical degree, in India.

In the absence of comparable surveys of the shareholders of other industrial companies it is impossible to say how typical the shareholders of C. Davidson & Sons Ltd. were. The most relevant general survey is Dr Cottrell's analysis of a 10 per cent random sample of newly formed English companies in 1860 and 1885. He found that in 1885 shareholders living within 50 miles of the company's registered office provided only 53 per cent of the share capital of these companies with more distant investors contributing the remaining 47 per cent. He also found that female investors and male skilled and semi skilled workers made a negligible contribution to their share capital whilst men engaged in trade and the professions and men not gainfully employed provided the bulk of the share capital.[6]

The share capital of C. Davidson & Sons was thus more local in origin despite substantial business elsewhere in Britain conducted through warehouses in Glasgow, Edinburgh, Liverpool, Newcastle on Tyne and above all London. Yet even in 1891, sixteen years after conversion of the business into a limited company and a succession of high dividend payments investors from Central and Southern Scotland and from the whole of England accounted for only 19 per cent of shareholders.

On the other hand women and male skilled and semi skilled workers were of much greater importance as shareholders than in the companies in Cottrell's sample in 1885. However the difference may be mainly a function of the greater age of the Scottish company, which attracted a rising proportion of more cautious investors, including ministers of religion as well as these two other groups, when the firm's proven track record had become apparent by the end of the 1880s.

Chapter 7

Owners and Directors

THE DAVIDSONS

For much of the 142 years of its independent existence C Davidson and Sons was a family firm, owned, controlled and run by members of the Davidson family. When Charles Davidson, the founder, died in 1843 his sons William and George, who had already assisted their father, took his place. A few years later George left to pursue a business career elsewhere leaving William in sole charge. William's five sons – Alexander, Charles, George, David and John – joined him in the business but he remained the sole proprietor until his death in 1873. However, even in those days of undiluted family ascendancy the necessity of obtaining a large loan from the North of Scotland Bank to finance expansion imposed some outside limits to freedom of action which was reflected in the regular inspection of the firm's accounts from 1858 onwards.

The conversion of the firm into a limited liability company in 1875, by enabling the investing public to buy shares in the firm, increased the possibility that one day ownership might pass from the Davidson family and that the directors, who were appointed by the shareholders and answerable to them, might be drawn partly or completely from outside the family. In fact the family retained 25 per cent of the issued shares of the company in their own hands at least until the early 1890s and when a major injection of capital was required in 1896 this was raised by the issue of debentures which posed no threat to family control provided that the stipulated interest was paid regularly.

The first directors were William Davidson's four surviving sons, the eldest having died in 1870. The death of George Davidson in April 1875 reduced their number to three and it was not until 1898

1 Mugiemoss in 1886

7 June 1811
Tack
Twixt
James Forbes of
Seaton
and
Charles Davidson

It is finally ended and agreed upon
between James Forbes of Seaton Heritable
Proprietor of the Lands of Mugiemoss
On the One part and Charles David
son Millwright presently residing
at Waukmill of Grandhome On the Other part as
follows viz.t The said James Forbes hath set and in
Tack and assedation lets to the said Charles Davidson
and his Heirs and Subtenants, with consent always
of the said James Forbes or his Successors first had
and obtained in writing, for the space of fifty seven
Years from and after the term of Martinmas last,
which is hereby declared to have been his entry
notwithstanding the date hereof, the Dwelling House
Barn and Byre on the Lands of Mugymoss, with the
yard belonging thereto, also part of the small park
or Inclosure at the back of the House and Garden of
Mugiemoss Also the ground along the brae between
the flourmill bed and Top of said Brae eastward
as far as the road from the mansion House of Mugie
Moss to the Spring well, and also the Haugh ground
betwixt the said Mill bed and the River from the ditch
at the head of the haugh downward to certain line

2 Lease of land at Mugiemoss by Charles Davidson, 7 June 1811

3 Mugiemoss Mills in 1953

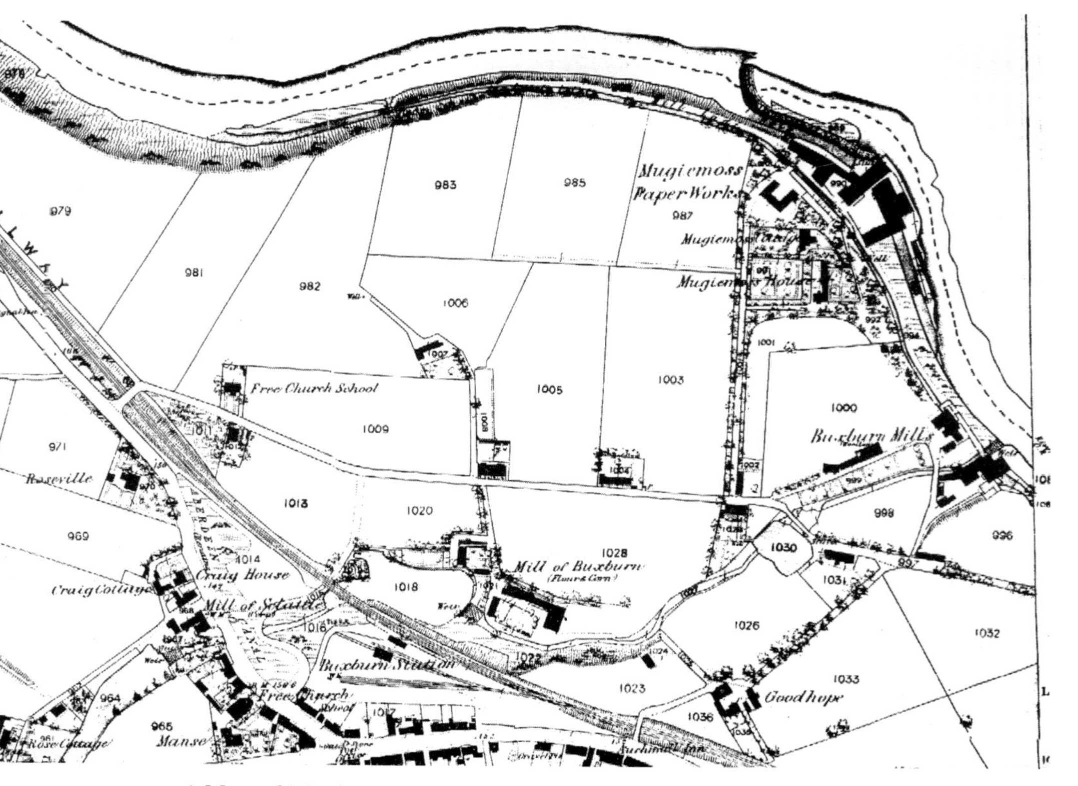

4 Map of Mugiemoss Mills and the surrounding area in 1867

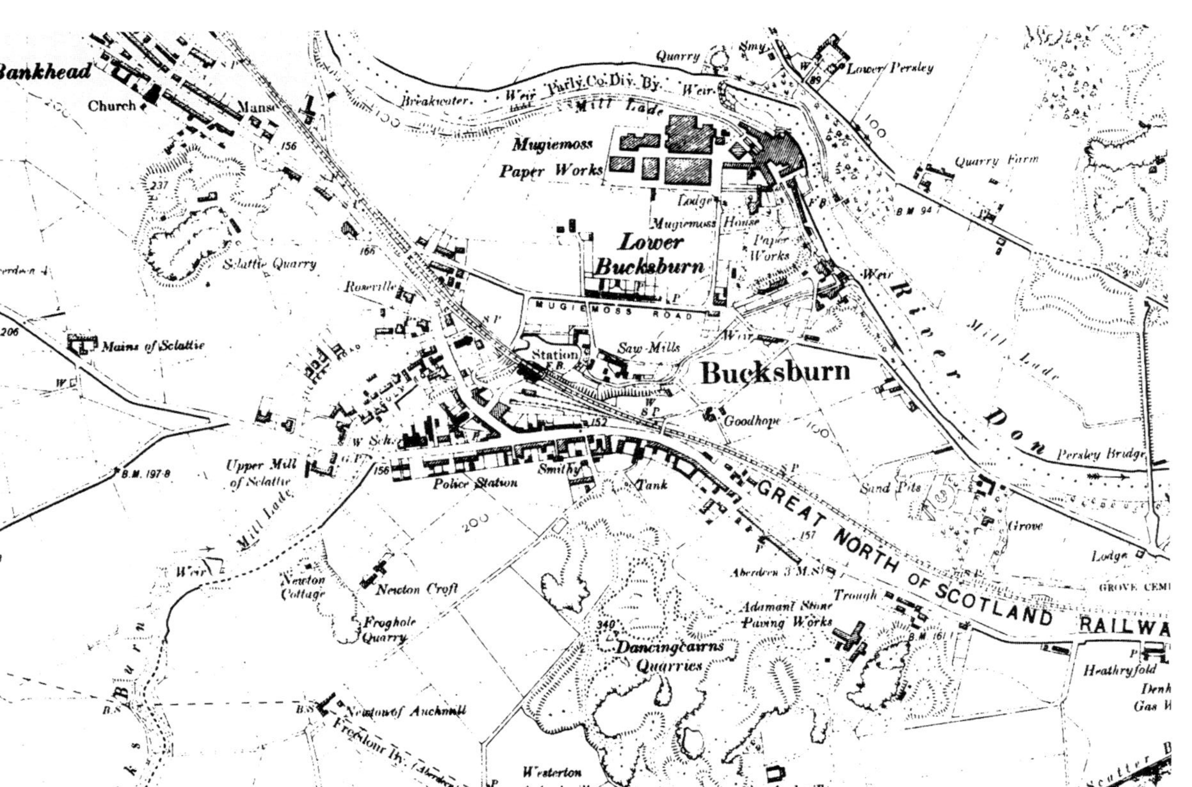

5 Map of Mugiemoss Mills and the surrounding area in 1902

PAPER MAKERS' ANNOUNCEMENTS. 73

MARK.

C. DAVIDSON & SONS, LTD.

Telegraphic Address— "PTARMIGAN, LONDON."

23, Upper Thames Street, LONDON.

ESTABLISHED 1796.

Mill 66-MUGIE MOSS, near Aberdeen.

Mill 80-BUXBURN, near Aberdeen.

LONDON: Paul's Pier Wharf, 23, Upper Thames St., E.C.
,, Size Yard, Whitechapel, E.
,, 62, Plumbers' Row, E.

LIVERPOOL: 14a, Peter's Lane.
NEWCASTLE: High Bridge.
GLASGOW: 59, 61 & 63, St. Enoch Sq.
EDINBURGH: Castle Terrace.

ABERDEEN: 4, Trinity Quay.

MANUFACTURERS OF

Middles, Browns,

PAPER MAKERS' WRAPPERS,

Grocery Papers,

CARTRIDGES, SMALL HANDS,

Caps, Thin Caps, Printings,

MANILLAS, AMMUNITION PAPERS, WATERPROOF PAPERS,

Glazed Papers of all Kinds,

PAPER FELT AND PAPER BAGS.

Printing of Paper Bags or Paper Wrappers done at the Mills.

Also

WHOLESALE WASTE PAPER DEALERS.

6 Advertisement for the firm's products in 1895

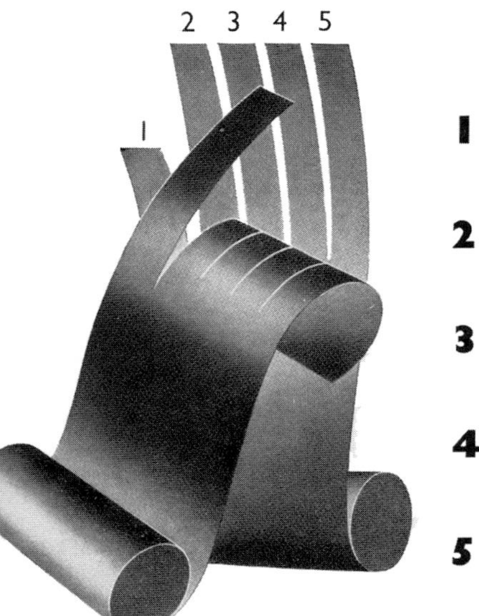

7 Advertisements for the firm's products in 1947. I Five worth remembering. II Variations on a paper theme

I Before alteration 3 March

II Trial run 28 April

8 Reconstruction of the Board Machine, Spring 1951

C. DAVIDSON & SONS, LTD.

MUGIE MOSS WORKS BUCKSBURN ABERDEENSHIRE

Telephone : Aberdeen 34

9 Indispensable Ibeco: Advertisement, 1939
I Ibeco concreting paper

Packed, cased, the products of Britain go out across the world, and British reputation with them. Major contribution to the safe arrival of both is IBECO — tough, adaptable, waterproof paper that provides sure protection for all damp-endangered merchandise. Extremes of climate do not affect it. Folding and creasing do not impair its waterproofness. It, too, is British.

COMPLETELY WATERPROOF KRAFT PACKAGING PAPER

9 1952. II Waterproof packaging to assist Britain's export drive
a. Dispatch from Britain

Between a product's reputation and the loss of it there's often no more than a thickness of paper. . . but the reputation's safe when the paper is IBECO. For a multivariety of merchandise leaving Britain for the four corners of the world, this tough, pliable, integrally-waterproof packing is making all the difference between "damaged in transit" and "received in sound condition" It's a good friend to Britain's good name.

IBECO

COMPLETELY WATERPROOF PACKAGING PAPER

MADE BY C. DAVIDSON & SONS LTD · MUGIEMOSS · ABERDEENSHIRE

9 II b. Arrival overseas

10 James Catto Duffus (1891–1962), chairman 1946–1953

11 David Peter Davidson (1910–1986), the last Davidson director

12 Richard S. Jukes, a director and later chairman of British Plaster Board Ltd, one of the architects of the 1953 merger

13 John Mackie, the first director not drawn from the family

14 Robert Watson McCrone, probably the most distinguished outside director before the merger

A.D. 1859, 1st DECEMBER. N° 2724.

Manufacture of Paper Bags, &c.

LETTERS PATENT to George Davidson, of Mugie Moss, in the County of Aberdeen, Paper Manufacturer, for the Invention of "IMPROVEMENTS IN PAPER BAGS, AND IN THE MACHINERY OR APPARATUS USED THEREIN."

Sealed the 7th February 1860, and dated the 1st December 1859.

PROVISIONAL SPECIFICATION left by the said George Davidson at the Office of the Commissioners of Patents, with his Petition, on the 1st December 1859.

I, GEORGE DAVIDSON, of Mugie Moss, in the County of Aberdeen, Paper
5 Manufacturer, do hereby declare the nature of the said Invention for "IMPROVEMENTS IN THE MANUFACTURE OF PAPER BAGS, AND IN THE MACHINERY OR APPARATUS USED THEREIN," to be as follows, that is to say:—

This Invention relates to the manufacture of paper bags according to an improved system either directly from the paper-making machine as is pre-
10 ferred, or from webs of reeled paper. In making the improved bags in connection with the paper-making machine, the web of paper as it exudes from the machine is severed longitudinally by rotatory cutters into the requisite widths. These severed widths then meet a set of creasing and cutting rolls. The roll which both creases the paper longitudinally, and cuts it transversely

15 George Davidson's first patent for making paper bags on power driven machinery, December 1859
I First page of the provisional specification

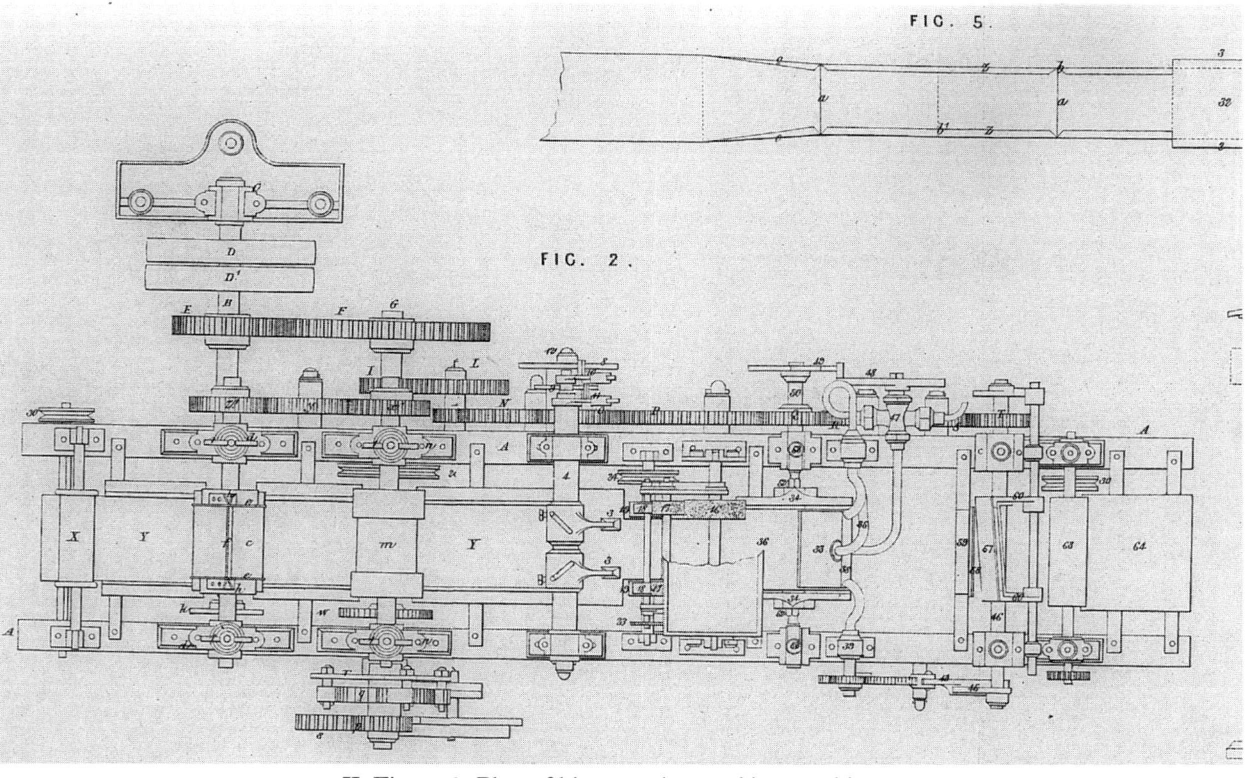

II Figure 2. Plan of his paper bag making machine

that additional directors were appointed, including the first person who was not a member of the family.

The death of John Davidson in 1897 precipitated the change. The following year Charles William Davidson, son of the Charles Davidson who died in 1870, and William Dalzell Davidson, the eldest son of Alexander Davidson, who had joined the firm in their youth, became directors. At the same time John Mackie, a long serving employee, was promoted, thus increasing the Board to the maximum of five permitted under the articles of association.

The upper limit to the number of directors was raised to six in 1908, in circumstances which are examined in section 2, to permit another person outside the family circle to be recruited. However after John Mackie's sudden death in 1911 the Davidson contingent was strengthened by the promotion of David Davidson's son Thomas, the future Colonel Davidson.

David Davidson's death at the age of 67 during the First World War left an unfilled vacancy but when Alexander Davidson died in London in 1920 after being knocked down by a car his third son Alan John Davidson became a director. His appointment marked a watershed, though one which was not apparent at the time, for in the following 32 years before the merger with British Plaster Board only one other member of the family joined the Board of Directors.

The dearth of Davidson directors reflected no failure of the male line but the severe economic difficulties facing the firm from 1922 to 1935, followed by four more years of uncertainty as sales and profits began to recover. On more than one occasion the business was close to failure and the mediocre profits generated were reflected in the total absence of share dividends for nearly 20 years. This was hardly the environment to encourage an ambitious son – or even a son-in-law or nephew – seeking a prosperous future, nor to reassure a less adventurous one wanting a safe and secure haven.[2]

In fact not merely were no other Davidsons appointed to the Board in the inter-war years but their number was depleted by the resignation of the two director sons of Alexander Davidson in 1926 after the firm's two Reserve Funds had been completely exhausted and the North of Scotland Bank had demanded extra security for the continuance of loans amounting to over £60,000. Colonel Tom and Charles William continued to struggle on assisted by a fresh outsider, A.T. Dawson, for the next nine years and after the reconstruction of the firm in 1935 they were joined by three other men drawn from outside the family, thus leaving the Davidson members for the first time, at least in numbers, in a definite minority on the Board.

During the war years and the post war years up to 1953 the Davidson presence was barely maintained. When Charles William died in 1943 Frank Williamson, an outsider, filled the vacancy created. The promotion of David Peter, Colonel Tom's only son, who joined the Board in 1946 after wartime service in North Africa and Italy, made some amends, but the two other appointments made at the same time were both from outside the family. He was in fact the last member of the family to be made a director of the firm and when his father died in 1951 the Davidson contingent was again reduced to a single director.[2]

The early education and training of the Davidson directors was rather narrow and inward looking. Education locally, in which Aberdeen Grammar School was prominent, and intensive work experience in the firm were the norm so that comparative technical and commercial data about the paper industry as a whole was drawn at second hand from the trade press, from informal contacts, and by recruiting foremen, managers and sales staff with previous experience in other paper firms. However, Alan John had several years' business experience in London and Charles William had worked for a firm of chartered accountants there before developing Davidson's sales as manager in turn of the Leeds, Edinburgh and London branches before returning to Edinburgh.[4]

Colonel Tom's early training was particularly impressive. He graduated in Chemistry and Engineering at Oxford University and after working 'right through the mills' at Mugiemoss he spent a year at Vienna Technical College, followed by a further year working in various paper mills in Austria.[5]

Whatever the lessons that had been learnt in the classroom or workplace the various sons did not have to face an interminable and frustrating delay growing crabbed and more conservative in outlook as they waited for promotion to the Board. All but one of the nine directors were relatively young men when promoted to the top level, with an average age of 31. The exception was Alan John who had pursued an independent business career in London and was 45 years old when he joined the Board in 1921 after the death of his father.[6]

George, the eldest of the first four directors in 1875, was 37,

David the youngest 27, but all four brothers had in fact been running the firm since their father was incapacitated in 1870. David Peter, the only other director appointed over the age of 35, owed his later advancement to several years of active service in the Second World War.

Wider experience of business and commerce generally was obtained through directorships in other limited companies, but almost invariably this followed after promotion to the Davidson Board. No fewer than seven of the directors were on the Boards of other companies, excluding firms controlled directly by C. Davidson & Sons. Locally based companies including Aberdeen Lime Co., Aberdeen Jute Co., Aberdeen Steam Navigation Co., Grove Cemetery Co. and Aberdeen Combworks Co., were to the fore and the typical holding, apart from companies in the Davidson stable, was two other directorships. However Alexander Davidson by 1910 was director of twelve other companies, on four of them as Chairman of the Board, which were primarily concerned with mining gold and other metals in the Empire and elsewhere abroad.[7]

What benefits were derived by C. Davidson & Sons as distinct from the individual directors involved is far from clear. No member of the family sat on the Board of other firms making, processing or selling paper products, except for the Davidson subsidiaries. Directorships of local shipping companies and the North of Scotland Bank may have conferred some commercial advantage but the benefits derived from a seat on the Board of Aberdeen Combworks Co. or Grove Cemetery Co. were less obvious, although certainly greater than any obtained from Colonial mining companies. No doubt decision making at the top level in other business concerns sharpened entrepreneurial wits but equally it may have dissipated energies and skills better channelled into re-shaping their own firm to meet fresh challenges.

The suspicion that the minds of the Davidson directors might have been focused more firmly on the family business in the late 19th and early 20th centuries when competition was growing more fierce and profits were declining is strengthened by a glance at their personal investments. On the death of John Davidson in 1897 the inventory of his personal or moveable estate and effects revealed that 60 per cent of the estimated value of £16,100 was held in shares and debentures of C. Davidson & Sons and sums owed him by the firm, whilst holdings in more than 20 other limited companies constituted a further 34 per cent of the total. Yet when David Davidson died in 1915 with a personal estate and effects valued at under £17,000 only 5 per cent of this represented holdings in the family firm. The value of shares held in three local firms alone – Aberdeen Dairy Co., Aberdeen Jute Co. and Aberdeen Lime Co. – exceeded the size of the Davidson holding, whilst the combined investment in four banking and insurance companies accounted for no less than 53 per cent of the total. Overall he was involved with 23 different concerns in addition to the family firm.[8]

A similar detailed picture is not available for Alexander Davidson who died in London in 1920. However the record number of directorships held, including the position of Chairman in four of them, suggests that he also was investing his accumulated funds in a variety of other businesses.[9]

In contrast to the lure of other industrial and commercial enterprises the Davidson directors, unlike some other leading British businessmen, found the attractions of a civic or political career easily resistible. It would appear that none were active in politics locally or nationally nor did they seek to serve the community on Aberdeen City or Aberdeenshire County Councils.

Colonel Tom was made a Justice of the Peace, his son in 1961 was appointed Deputy Lieutenant of Aberdeenshire, and David was a life governor of Aberdeen Royal Infirmary. Otherwise, apart from membership of local churches, participation in the life of the community beyond Mugiemoss seems to have been confined to sporting activities.[10]

Alexander Davidson was a keen golfer, a pillar of the Wimbledon and Sandwich Clubs. He was an expert shot, competing in National Rifle Association meetings in his younger days, and was also President of Wimbledon Curling Club. Colonel Tom devoted his leisure time to the traditional pursuits of country gentlemen – hunting, shooting and fishing – but in his youth he was also an accomplished rugby player and appeared in several trials held by the Scottish Rugby Union.[11]

However in one other sphere two of the directors gained public attention and esteem which for some time must have been the envy of many local councillors and politicians. Both colonel Tom and his only son David Peter were war heroes.

Thomas Davidson was commissioned in the Territorial Army in the Royal Artillery in 1900. During the First World War, serving in

France, he was awarded the D.S.O. and Croix de Guerre with palm and was mentioned in Haigh's dispatches on seven occasions, ultimately commanding the 255th Brigade Royal Field Artillery. His son served in North Africa and Italy in the Second World War and in 1945 was awarded the Military Cross.[12]

It is difficult not to see these years spent in military conflict, however valuable to the nation's war effort, as a waste of scarce entrepreneurial skills which might have been employed in the service of the family firm. Both men already had experience of leadership at top level in the firm, the son as a manager and director of the subsidiary company Davidson's Paper Sales Ltd., before their wartime careers began.

However the reputation gained by Colonel Tom in the First World War may have yielded some positive benefits in the troubled inter-war years. The respect and admiration in which he was held by many of the workforce may have helped with the hard tasks of maintaining morale and ensuring the discipline needed to minimise the risk of fire, to eliminate waste, and to economise generally. Similarly local creditors, ranging from the North of Scotland Bank to all those concerned in the capital reconstruction of 1935, may have been more tolerant of the firm's problems out of sympathy for the Colonel, feelings of guilt if he were ruined and a wish to avoid a wave of adverse publicity.

THE OTHER DIRECTORS

The fourteen directors recruited from outside the ranks of the Davidson family joined the Board when they were somewhat older than the Davidson sons. The oldest director appointed was 63 and their average age was 46. Significantly the youngest, James Mearns Dawson, aged 32, was the only one who was the son of an existing director in the firm and his previous experience of management at the top level in the wider world of industry and commerce appears to have been confined to Morrison's Economic Stores, the Aberdeen retailing business in which his father was also a director.[13]

Four of the directors were given their seat on the Board as a recognition of the prominent part they already played in running the firm. John Mackie, the very first outsider to be promoted, who became a director at the age of 49 in 1898, was the archetypal company servant, rewarded for long and faithful service. He was born in Aberdeenshire in 1849 and began work at Mugiemoss in his youth. He joined the London Office in 1871, was one of the earliest shareholders when the firm became a limited company, soon owning 200 shares of £1, and by 1893 was manager of that office with the handsome salary of £600 a year.[14]

Frank Williamson on the other hand was in his late thirties when he joined the firm. He had gained substantial experience of paper board production working for other firms in the industry and was recruited to introduce the manufacture of boards at Mugiemoss as part of the programme of diversifying output begun after the reconstruction of the company in 1935. Within four years of his appointment board manufacture had outstripped paper production and had become of major importance at Mugiemoss. His achievements were recognised in 1943 when he was given a seat on the Board and made joint Managing Director.[15]

James Partington had been company secretary for two years when he joined the Board at the age of 37 in 1948, having previously been employed by a bus company, Lancashire United Transport. However E.J. Warburton, who had wide experience of paper and board manufacture at home and abroad, including a lengthy spell with St Anne's Board Mill near Bristol which was a major producer of boards, was promoted within a year of his appointment as mill general manager at the age of 45.[16]

Only one other director, Alexander Marr, brought to the Board first hand experience of the paper industry. He was 63 when appointed and was already a director of the neighbouring firm of Alexander Pirie and Sons which was then probably the largest paper-making firm in Scotland. He had become a director of Pirie's four years earlier in 1904 after 35 years' service with the firm, including 22 years as secretary. The maximum size of the Davidson Board as we saw earlier, had been increased to make possible his recruitment and the appointment was made at a time of crisis when, for the first time since the company had been constituted, shareholders received no dividends.[17]

Six of the remaining eight directors brought to the Boardroom a wide range of industrial and commercial experience, derived in part at least from some years as directors of other firms before joining C. Davidson & Sons. The most outstanding, whose achievements were deemed worthy of an obituary in the *Times*, was Robert

Watson McCrone. He was a pupil at Merchiston Castle School, Edinburgh, and after a scientific education at the Royal College of Science and Technology and at Glasgow University where he was awarded a degree in engineering he served as an officer in the Royal Engineers throughout the First World War. Subsequently he launched his successful career by founding Metal Industries Ltd of which he was Managing Director from 1923 to 1951. He was also a pioneer of the distribution of liquid oxygen, becoming a director of British Oxygen Ltd in the early 1930s. Among his notable successes in these early years, springing from these two interests, was the raising and dismantling of a large part of the German Navy scuttled at Scapa Flow at the end of the war.

He joined the Davidson Board in 1936 after the reconstruction of the company had been agreed with creditors and the appointment was of precisely the kind to re-assure the new debenture holders that the firm was in experienced and capable hands. He resigned in 1946 after all the debentures had been repaid or converted into ordinary shares to pursue his varied interests as director of a number of other companies, until late in life he tired of industry and turned to farming.[18]

J.A. Montgomerie who became a director at the age of 58 at the same time as McCrone, also had business interests outside the North East of Scotland and his appointment was similarly designed to re-assure nervous debenture holders. The son of an Ayrshire cattle breeder, he began his working life as an analytical chemist employed by the Glasgow firm of paint manufacturers, Thomas Hinshelwood & Co. Ltd. He founded his own paint manufacturing firm, Montgomerie, Stobo & Co. in Glasgow in 1906 and became its managing director. However there were also more tangible reasons for his appointment to the Davidson Board since he was a director of British Bitumen Emulsions Ltd and International Bitumen Emulsions Ltd the firm which had licensed Davidsons to use the new patent Ibeco waterproofing process which was to prove so successful.[19]

The four other businessmen who became directors were local men whose enterprises had their headquarters in the Aberdeen area. W.R.S. Mellis, appointed in 1946, after education at Tonbridge School, Kent, and Cambridge University, had begun work in George Mellis & Son, the family firm of wholesale merchants founded by his great grandfather in 1844, eventually becoming

Chairman after its conversion to a limited company. During the 1930s he also joined the Board of Mitchell & Muil Ltd, wholesale and retail bakers and biscuit manufacturers, and was the Managing Director when he was given a seat on the Davidson Board.[20]

H.R. Spence had lengthy experience of textile manufacture. He was the son of a hosiery manufacturer and was the managing partner of the family firm of William Spence & Son based at Huntly, Aberdeenshire, as well as the managing director of another hosiery firm, Harrott & Co. Ltd. of Aberdeen when he became a director of Davidsons in 1951 at the age of 58. However his appointment probably owed more to the fact that he was a local M.P., elected for Aberdeenshire West in 1945, rather than to any desire to employ his business skills.[21]

B.W. Tawse who became a director in the same year controlled a firm which operated far beyond the city boundaries. In 1940 he took over as Chairman of William Tawse Ltd, the public works contractor and civil engineering firm started by his grandfather at Kemnay, Aberdeenshire at the end of the 19th century. The firm retained its headquarters in Aberdeen but had expanded its operations throughout Scotland, including among its major activities the building of air fields for the Ministry of Defence during the war and a series of Hydro Electric power schemes in the post-war years. Like Spence the prestige his name brought to Davidsons was perhaps as valuable as his business skills although the prospect of additional orders for Ibeco paper may have been a further consideration.[22]

A.T. Dawson's place in the history of C. Davidson & Sons was more crucial. He joined the Board in 1926 after Alexander Marr's health had deteriorated and Alexander Davidson's sons, Alan John and William Dalzell, had also resigned as the two remaining directors began the long uphill struggle to keep the firm alive. The financial assistance given by himself and other members of his family played an important part in the firm's survival and by January 1938, as can be seen in Appendix B, he held nearly half the issued share capital. When he retired in 1946 the future of the firm was secure and prosperous.

He was born at Buckie, some 72 miles North West of Aberdeen in 1885, the son of a grain merchant and distiller. In his youth he was a wine merchant in Glasgow but after his marriage to Isobel Mearns of Aberdeen in 1910 he moved to Aberdeen to begin a new career at Morrison's Economic Stores, a firm retailing cheap clothing, known popularly as 'Raggy Morrisons'. Isobel's father, James Mearns, owned the stores and in 1924 when the firm became a limited company Dawson was one of the three directors and held approximately one third of the equity.[23]

Two Aberdeen lawyers complete the list of directors. J.C. Duffus, who joined the Board in 1936 after the re-construction of the company, was an experienced advocate. The son of an Aberdeen advocate he had become a partner in the family legal business of Wilsone and Duffus in 1920. He became Chairman in 1946 presiding over the ambitious post-war scheme of expansion and was the main architect of the plans to merge the firm with British Plaster Board.[24]

The career of R.M. Ledingham, who was 18 months younger, followed a similar pattern in his earlier years. He also was educated at Aberdeen Grammar School and Aberdeen University where he graduated M.A. in 1913 and after war service Ll.B. in 1920 before joining his father's legal firm, Edmonds and Ledingham. Both men were keen members of the Grammar School Former Pupil's Club. The offices of the two legal firms were situated within easy walking distance of each other in Golden Square, Aberdeen and there seems little doubt that both men were well acquainted if not close friends. When Ledingham became a director in December 1951 at the age of 58, Duffus was already Chairman and the appointment strengthened Duffus' hand in planning the firm's future including the negotiations which led to the merger in 1953.[25]

The enthusiasm for the world of industry and commerce shown by the 14 directors, like those drawn from the Davidson family, was scarcely dimmed by the counter attractions of political or civic activity. J.A. Montgomerie was a keen supporter of the Unionist cause in Cambuslang and became president of the local association, but apart from H.R. Spence, who entered Parliament at the age of 48, it would appear that the directors did not seek civic or political office.[26]

Other forms of public service, if they were undertaken at all, were on a relatively modest scale. W.R.S. Mellis was a J.P., Deputy Lieutenant of Aberdeen, a trustee of Aberdeen Savings Bank and a member of Aberdeen Harbour Board. R.W. McCrone became a governor of the Royal College of Science and Technology, Glasgow and was Chairman of the Dunfermline and West Fife Hospital

Management Committee in pre National Health Service days, donating a house and grounds to the local hospital as well as one of the houses for disabled ex-servicemen at Cherrybank, Dunfermline. J.C. Duffus also devoted time and energy to hospital administration, serving for many years on the Board of Management of Aberdeen General Hospitals, part of this as Chairman.[27]

Sporting activities provided another outlet for surplus energies and some further contribution to the life of the local community. Mellis, who was awarded a blue for badminton at Cambridge University, and Montgomerie and Tawse were keen golfers. Duffus played cricket for Aberdeen University and Aberdeenshire and in his younger days played rugby as did Ledingham who became President of the Scottish Rugby Union. Spence and McCrone in his earlier years were yachtsmen, but Spence, who also enjoyed golf and shooting, excelled at ski-ing, winning the British Cross Country Ski Championship at Murren in 1929.[28]

The careers of several future directors were interrupted by military service in the First World War. In this their experience was no different from that of many fellow countrymen. However two had distinguished records which may have had some bearing on their future links with Davidsons. Duffus and McCrone both served as officers in France and both were awarded the Military Cross and Croix de Guerre. They thus shared a common experience with Colonel Tom Davidson and it was therefore singularly appropriate that both should join the Davidson Board in 1936 to assist the Colonel in his efforts to revive the firm after its rescue from near oblivion.[29]

Overall the directors drawn from outside the Davidson family brought to the boardroom experience of a kind which was in short supply among the Davidson sons, namely first hand experience of other firms in the paper industry and top level experience of other sectors of industry and commerce including hosiery manufacture, food production, construction and textile retailing. This fund of direct experience increased from the mid 1920s onwards as the Davidson domination of the boardroom began to wane, although Colonel Tom Davidson was in fact the best equipped of the Davidson sons with a scientific and technical education at Oxford University and Vienna Technical College, followed by work experience in Austrian paper mills.

However like the Davidson sons the group's business skills and expertise were gained essentially through practical work experience rather than by formal education and training. Their technical qualifications, McCrone excepted, were negligible and only five were University graduates. But in this respect they were no different from the majority of Scottish businessmen whose biographies appear in the Dictionary of Scottish Business Biography published between 1986 and 1990. As Professor Slaven, one of the editors, observed, 'gaining experience of the business through training on the job remained the norm for most proprietors of Scottish business'.[30]

Chapter 8

The Labour Force

PAPER MANUFACTURE AT MUGIEMOSS

The Basic Raw Materials

The manufacture of paper formed the core of business operations at Mugiemoss for more than a century until the late 1930s. Millboard production was then introduced and rapidly assumed a greater importance.

In the 1820s when Charles Davidson began paper-making on the site, the paper industry was still dependent on the traditional fibrous raw materials – waste linen and cotton rags, supplemented by old ropes, sails, and twine together with waste paper clippings. However by the middle of the 19th century as a result of the continued expansion of the industry, supplies of old rags proved inadequate to meet the growing demand. Continental supplies were increasingly used by European and American paper manufacturers, whilst in Britain wholesale rag merchants, who were dependent on the efforts of itinerant rag-and-bone men and the willingness of numerous individual households to save their old rags, found it difficult to increase supplies. As Professor Coleman observed

> An increase in rag prices would not readily call forth an appreciably greater quantity of a waste product thus collected. The rag merchant, sitting astride this inelastic supply line, was not likely, nor indeed would he have found it easy, so to re-organise methods of collection as to produce a substantially larger flow of rags.[1]

In response to this situation the use of esparto grass, imported from Spain and North Africa, spread rapidly in the paper industry from the 1860s onwards. Imports exceeded 95,000 tons a year by

1868 and by the 1880s the annual import was more than twice this figure. Esparto proved especially popular with Scottish producers and soon constituted a major part of the input of Scottish paper mills, particularly those which concentrated on the production of writing and printing papers.[2]

The pressure on raw material supplies was eased still further by growing imports of wood pulp. Wood pulp produced mechanically was of relatively poor quality and therefore of limited appeal to most British paper manufacturers except those making cheap quality paper such as newsprint. But once technical advances made possible the production of high quality wood pulp by chemical processes, its use spread rapidly in the United Kingdom. Wood pulp imports were negligible before 1870, amounting perhaps to some 2,000 tons by 1869, but ten years later the combined import of wood pulp and rag pulp was nearly 28,000 tons and by 1887, when the customs enumerators at last distinguished wood pulp imports separately from those of rag pulp and miscellaneous vegetable fibres other than esparto, more than 79,000 tons were imported, rising to nearly 138,000 tons in 1890.[3]

However Mugiemoss and some other paper mills continued to use substantial quantities of rags until at least the beginning of the twentieth century since their pulped fibres provided the toughness and resilience which were desirable in certain types of paper such as bank notes, ledgers and account books, and the better quality wrapping paper. Moreover, the widespread switch of paper manufacturers to esparto and wood pulp meant that rag prices declined considerably from the peak figure reached at the end of the 1850s and except for some brief intervals continued to fall until the end of the century, thus encouraging these mills to continue using rags as a source of fibrous raw material.[4]

At the beginning of the 1870s when the royal commission of enquiry was investigating the pollution of Scotland's rivers by industrial and other effluent, Mugiemoss was consuming some 3,000 tons of rags per year and a mere 5 tons of esparto.[5] There are no comparable figures for the remainder of the 19th century but the continuing predominance of rags as a raw material is not in doubt. It is significant that when steps were taken to increase the output of paper shortly before the turn of the century the investment programme included the purchase of 8 new rag engines for the conversion of rags into pulp and a factory inspector visiting

Mugiemoss in 1901 commented on the large number of women employed in the Rag Sorting Department.[6]

Both waste paper and wood pulp were also employed as paper-making materials at Mugiemoss at various times in the 19th century although to what extent is uncertain. They were used in making the heavier coarser brown papers which formed only a part of the firm's output and interest in these fibrous raw materials was influenced by the current price of rags, increasing substantially for example at the time of the Franco-Prussian war when European rag prices for some months rose sharply in price.[7]

George Davidson, one of William Davidson's sons, built an experimental plant to make mechanical wood pulp about 1867 but the project was soon abandoned. The Minute Book recording the decisions taken by the Davidson brothers after their father became incapacitated noted in August 1871 'That wood be laid aside for the present so far as moderate experimenting goes' and in March 1872 when the next set of Accounts was being considered it was confirmed that 'Almost nothing has been done in experimenting on wood'.[8]

However the provision of supplies of waste paper and wood pulp in the early 1870s left a more permanent legacy since it was decided to lease a warehouse in London for the collection and storage of waste paper and other fibrous raw materials, both for the firm's own needs and for re-sale to other users. Size Yard, Whitechapel, was leased for this purpose at a rent of £100 a year in June 1872. It remained in use for over 20 years, despite a disastrous fire in its first year of operation, and advertisements by the firm in the 1890s included the description 'Wholesale Waste Paper Dealers' as well as listing the main products manufactured at Mugiemoss.[9]

A decisive shift away from rags as the main fibrous paper making material did not occur until the first decade of this century. It was argued in Chapter 3 that the financial difficulties caused by mounting foreign competition in the production of the traditional heavy wrapping papers which formed part of the output of Mugiemoss led to the decision to install a wide MG Paper-making Machine. This increased the output of glazed papers, which were thinner lightweight papers combining toughness with flexibility, and made it possible to introduce new products to the firm's range made from chemically treated wood pulp, including the outstanding Kraft Paper which originated in Sweden soon after 1900.

When the First World War broke out in 1914 there were large stocks of wood pulp at Mugiemoss and the difficulties in maintaining paper output at the plant two years later were attributed by the directors, labour shortages excepted, to the virtual exhaustion of these stocks and the failure to secure further supplies of wood pulp "of which the company is a large consumer'. Concern about long term supplies after the end of the war also led the directors, like those of some other firms in the industry, to make a substantial investment in the Bay Sulphite Company Ltd. of Canada which operated wood pulp mills at Ha! Ha! Bay.[10]

In the late 1930s however, the composition of the fibrous raw materials used at Mugiemoss was again transformed. The newly installed Board Machine, which soon produced a tonnage of millboard exceeding the output of paper on the existing paper-making machines, consumed pulp made from waste paper and this raw material now became of major importance.[11]

Conversion to Paper

The first stage in the manufacture of paper was conversion of the fibrous raw materials into pulp ready to flow onto the paper-making machine. Wood pulp purchased from outside suppliers required the least preparation at the mill, rags the most.

Rag wholesalers normally sorted and graded rags into one of some twenty to thirty categories each with a different price. But on arrival at the paper mill further sorting was required. Some dust was first removed in a revolving hollow cylinder covered with very coarse wire and then the rags were cut into pieces of uniform size, the undesirable parts such as silk, elastic, and buttons were removed, and the seams were cut open. The cutting usually took place on a bench with a top consisting of wire netting through which dust escaped. However, the dust remaining was loosened and removed subsequently by subjecting the rags to violent agitation in a willow or dusting machine.

The rags were then taken to a spherical rotary Rag Boiler. A solution of caustic soda was added and steam passed in whilst the boiler rotated slowly for about eight hours. When the dirt in the rags had been thoroughly loosened the boiler was stopped, the dirty water was run off, and clean water was run in, the boiler being then revolved to rinse the rags.[12]

In the next stage of treatment the rags were reduced to smaller fragments and then to a mass of threads known as half-stuff. This took place in a breaking engine where a heavy revolving roll with knives projecting from it macerated the rags between the roll and the bed plate of the engine on which there were also projecting knives. In the process more dirt came away, a hollow drum in the engine covered with wire gauze acting as a washing drum through which the dirty water passed and was then taken away.

Manufacturers of writing and printing papers and other kinds where the final colour was important would then bleach the half-stuff in a weak solution of bleaching powder to remove the brownish tinge. However in the production of most wrapping and packing papers this was not necessary.

The half stuff was converted to a fine pulp in a beating engine or beater, which was similar in construction and function to the rag breaking engine, although the revolving roll was capable of finer control as the distance between the roll and the bed plate was adjusted during the beating operations. The similarity of both types of engine meant that the beater was often used to perform both processes – the initial disintegration of the rags into half-stuff and the final conversion to pulp.

The texture of the pulp varied, to suit the type of paper being made, according to the length of time taken, the sharpness of the knives, and the rate at which the beater roll was lowered in the mixture towards the bed plate. It was not unusual to blend together and process in the beater, the pulps of two or more different fibrous materials such as rags and wood.[13]

China clay or some other loading material was also added and blended into the mixture. This filled up the pores of the paper, giving the sheet a closer texture with a smoother surface which made it more receptive to printer's ink and also facilitated subsequent finishing processes such as calendering.

Rosin or some other sizing agent was a further vital ingredient added in the beater, accompanied by alum to fix the size to the pulped fibres. Size would not be used in making blotting paper or other papers where absorbency was an essential characteristic but for other types of paper it helped to provide some degree of resistance to penetration by water and by printer's ink.[14]

If coloured paper was being made the appropriate dyestuffs would be added to the beater and blended into the mixture. Here too, alum was used as a fixative agent.

Wood pulp needed little treatment before it went to the beater. It usually arrived at the mill in the form of pulp boards which were readily broken up and converted to half-stuff. The treatment of waste paper when it had been sorted and graded was equally simple, the paper being steamed and then pulped.[15]

The finished pulp flowed from the stuff chests in which it was temporarily stored through sand traps and strainers which intercepted grit, unbeaten particles, and knots of fibre, onto the endless band of woven wire supported on a series of rollers which formed the heart of the Fourdrinier Paper-making Machine. The surplus water fell through the meshes of the wire and more was extracted by the open mouth of pumps in suction boxes beneath the wire, the water being re-used in future batches of pulp. Meanwhile, the pulp settled down on the surface of the wire in the form of a wet sheet.[16]

The wire of the paper-making machine was some 50 or more feet in length and its width ranged from 48 inches or less for an older machine with a relatively modest output to 150 inches or more for a modern, far more expensive, machine with a substantially greater output. The widest paper-making machines were best suited to long production runs of a standard product in continuous demand such as newsprint. The two Fourdrinier machines at Mugiemoss employed wire 63 inches wide, the single machine at Waterton Mill which was taken out of production by the beginning of the 1880s was a 54 inch machine.[17]

The sheet of paper was carried by the wire through the couch rolls where the paper was couched or pressed by a felt covered roll to consolidate the paper, the wire returning to the beginning of the machine to receive more pulp. Meanwhile the limp paper was carried forward to the press rolls where it was further pressed by polished rollers to remove the wire and felt marks.

The paper then travelled over a series of large hollow cast iron drying cylinders which were heated internally by the exhaust steam from the engine driving the paper-making machine. The heat and speed were carefully controlled to prevent the paper being spoilt by too rapid drying.[18]

At the end of the Fourdrinier paper-making machine there were stacks of rolls through which the paper was led to give the final finish to the surface of the paper. However, as was noted in Section 1 of Chapter 2 much of the paper at Mugiemoss was not 'machine finished' in this way but was given a glazed surface in a separate operation after leaving the drying cylinders, either by passing the paper through a super calender or through a friction calender.[19]

In the MG paper-making machine installed at Mugiemoss in 1909, which produced paper 108 inches wide, the glazing of the paper, as we saw in Chapter 3, was done on the machine itself. The wet end of the machine was similar to that of the normal Fourdrinier machine but at the dry end a single large drying cylinder replaced the series of drying cylinders on the Fourdrinier and as the web of paper passed over the single cylinder it was dried and glazed on one side, hence the term machine glazed (MG).

Some papers required somewhat different treatment. At Mugiemoss, paper felts of various kinds including cedar felt and waterproof felt, whose production was discussed in some detail in Chapter 2, came into this category, as did pasteboard, high quality surface card used for menus and visiting cards, which was made by pasting good quality paper onto both sides of an inferior *middle*.

The final stages of production in the typical paper mill involved cutting the reels of machine finished and of glazed paper into smaller widths and then single sheets on simple machines. After this the sheets were taken to the *salle*, the sorting and packing room, where the sheets were quickly examined for defects by hand, trimmed and cut into smaller sheets if necessary, counted, and wrapped into parcels which were tied with string or tape ready for dispatch to the customer.[20]

However at Mugiemoss a substantial proportion of the paper was not supplied to the customer in the form of paper sheets but as printed bags. Bags of all shapes and sizes were made on a variety of machines, some driven by power but others, performing operations which were not easily or economically mechanised, still worked in the early years of this century by hand. Some of the bags would be dispatched with a plain surface but increasingly they were printed at the mill with names, addresses and other advertising details to meet the customer's specification.

THE SIZE OF THE LABOUR FORCE

The number of workers employed by Charles Davidson, the founder of the firm, is uncertain but it is unlikely ever to have exceeded one hundred. Two years after his death in 1843 seventeen of the larger paper mills in Scotland employed an average of 65 persons each and the four papermaking mills in the Aberdeen area had a combined labour force of some 300–400 persons, of whom 150 were employed by Alexander Pirie & Sons, the largest firm.[21]

By the beginning of the 1870s after a period of sustained expansion the labour force had grown considerably. The Royal Commission investigating the extent of river pollution in Scotland which collected data on mills beside the river Don in the course of its enquiries recorded a total of 216 hands employed at Mugiemoss.[22]

Over the next twenty five years, further growth in output and diversification into new products generated a substantial expansion of the labour force. In 1896 when the firm was celebrating the centenary of Charles Davidson launching his first paper-making enterprise, the *Aberdeen Journal* reported that approximately 600 work people were currently employed. A contributor to the biographical survey of some leading Scottish businessmen, *Men of the Period: Scotland*, published soon after this, claimed that the Davidson directors 'now employ upwards of seven hundred hands'.[23]

Neither survey indicated whether the figure included clerks, warehousemen and commercial travellers employed at the warehouses in London, Liverpool, Newcastle on Tyne, Glasgow and Edinburgh. The lower figure given in the *Aberdeen Journal* may possibly be due to their exclusion.

A dozen years later when sales were stagnating and profits had declined in a harsher economic climate the size of the labour force had probably passed its peak. A survey listing the numbers employed in the main departments of the firm drawn up in June 1908 as the basis for an insurance contract to cover the firm's liability against possible claims under the tougher Workmen's Compensation Act of 1906 gave a combined total of 674 employees including 29 travellers and 63 clerks.[24]

In the years between the two world wars as the firm struggled to survive, the labour force was substantially smaller. The sole surviving wages book records 365 workers at Mugiemoss in February 1930 and more than six years later following the capital reconstruction of the company, the *Press & Journal* in an interview with Colonel Davidson reported that about 350 persons were employed at the Works, a figure which he anticipated would rise to more than 400 when the new Board Machine was in full operation.[25]

The programme of expansion launched after the end of the

Second World War transformed the company's position as an employer. The modernisation of the Mugiemoss mills ensured the preservation of jobs locally and also created new openings but the post war constraints upon expansion examined in Chapter 5 meant that the output of paper and board at Mugiemoss remained relatively static until 1950. Meanwhile the acquisition of three English firms – at Peckham, Radcliffe, and Gateshead – between 1945 and 1948 – added several hundred workers to the payroll and made C. Davidson & Sons the parent firm in a group of companies which produced paper and millboard and converted it into a variety of products. By the beginning of 1950 the directors were responsible for the employment of some 1,050 persons, the largest total in the firm's history. But although Mugiemoss remained the Head Quarters of the group a substantial proportion of the workforce, probably in excess of 50 per cent, were employed at sites many miles distant from Aberdeen.[26]

OCCUPATIONS AND PAY

Information about the composition of the labour force at Mugiemoss is even more sparse. The number of persons working in each of the main departments of the mill is known only for two periods, in 1908 and in 1930. However it seems likely that once the firm had embarked upon the production of printed paper bags on a substantial scale in the final quarter of the nineteenth century, although paper manufacture continued to form the core of operations, a large part of the labour force was not directly involved in its production.

Of the 674 persons employed in 1908 the papermaking department provided work for 105 of them and another 34 were preparing rags for conversion into paper. However even if it is assumed that all the 60 persons described as warehousemen and girls were engaged in finishing processes such as cutting, sorting and packing paper, the combined employment in papermaking would be less than 200 – a mere 30 per cent of those on the payroll. On the other hand the machine and hand bag making departments employed 264 persons – 39 per cent of the labour force – whilst 29 engineering workers serviced the machinery throughout the mill.[27]

The provision of office staff and sales personnel, who comprised 63 clerks and 29 commercial travellers, seems reasonably generous by the standards of the time, perhaps reflecting the fact that the firm was selling the products of other paper firms as well as its own manufactures. Two of the three main carpet manufacturing firms, Templetons of Glasgow and Tomkinson & Adam of Kidderminster, in 1900 for example respectively employed 85 office staff out of a workforce of more than 2000 and 65 office workers out of 1,200 employees, whilst in industry as a whole at the time of the 1911 Census clerical workers, including foremen and supervisors, accounted for less than 8 per cent of the total labour force.[28]

In 1930 data on the number of office staff and sales personnel has not survived. Of the other 365 workers employed, 145 were engaged in the hand and machine bag making departments, almost 40 per cent of the total, and 27 were in the Printing Department. The preparation of pulp and its conversion into paper employed 102 persons and 53 were employed in various finishing processes, giving a combined total of 42 per cent of the workforce directly involved in papermaking. The number of engineering workers meanwhile had increased from 29 to 30 despite the reduction in the size of the labour force since 1908.[29]

Women and girls constituted a substantial proportion of the workers employed in the paper mill but they were concentrated in the paperbag making department and in the preparatory and finishing processes of papermaking, principally in sorting and cutting rags and in inspecting the finished paper for faults. Printers, engineers, managers and foremen were exclusively male and the main papermaking operations including rag boiling, beating the pulp and running the papermaking machine, were also a male preserve whilst in the finishing processes men worked as reelermen, cuttermen and tiers and packers of the finished product.[30]

The exact proportion of female workers employed at Mugiemoss canot be determined, but it is probable that they constituted substantially more than half the workforce towards the end of the nineteenth century. The Census returns for Aberdeenshire, which recorded the paper workers employed by Culter Mills Paper Co, Alexander Pirie & Sons of Stoneywood, Thomas Tait and Sons of Inverurie and the struggling Gordon's Mills Paper Co, as well as the workers at Mugiemoss, reveal that in the years 1871 and 1891 female workers comprised some 57–58 per cent of the labour force engaged in paper manufacture whilst in 1881 the proportion was almost 67 per cent.[31]

The concentration of Alexander Pirie & Sons on high quality papers which required a substantial proportion of workers, many of them female, to be employed in the finishing processes may have boosted its share of the Aberdeenshire total at the expense of the other three firms. On the other hand in the case of Mugiemoss this bias was offset by the omission of paper bag makers, who were classified with paper box makers and recorded in the printed census tables separately from persons directly involved in paper manufacture. In 1891, for example, in addition to 1094 females recorded in paper manufacture in Aberdeenshire, another 333 females, together with 7 males, were recorded in paper box and paper bag making and it seems likely that the bulk of these persons were employed by C Davidson & Sons.[32]

A variety of skills were required in paper manufacture. The most highly skilled and best paid workers were the men at the very heart of the papermaking operations, the beatermen making the pulp and the machinemen operating the papermaking machine, who with their assistants generally comprised about 10 per cent of the labour force in a paper mill.[33]

The beaterman carefully adjusted the composition of the pulp in the beater to suit the particular type of paper being made. He would decide from experience when the pulp had reached the correct texture, testing it by hand to see if it 'felt right'. He was always on the look out for impurities which might spoil the finished paper, such as undissolved grains of dyes and unbeaten or undyed stuff which had stuck to the cover or sides of the beater and had been washed subsequently into the stuff chest. On his skill and judgement in producing suitable pulp the strength of the sheet of paper laid down on the papermaking machine, its closeness, its colour and its surface characteristics ultimately depended.[34]

The competence and experience of the machineman was no less important. It was essential to keep the papermaking machine running smoothly at a constant speed in order to prevent variations in the thickness of the finished paper, a task which was not easy when steam engines and rope drives provided the driving force in the nineteenth century. Even a slight variation in the steam pressure from the boilers affected the speed of the machine and required prompt adjustment.

Wires, felts and rolls had to be kept clean from impurities which might create faults in the sheet of paper and it was necessary to maintain an even flow of pulp on the wire to ensure an even sheet of paper, unspoilt by thick and thin places. The machineman had also to regulate the inflow of steam into the drying cylinders controlling the heat and speed to prevent hasty drying of the paper, which would result in defects such as cockling, wavy edges, excessive thinning, and even breaking of the web.[35]

Adjustments to the machine, on which the wire belt was typically moving at 250 feet per minute at the turn of the nineteenth century, were not made any easier by the economic imperative to keep the papermaking machine in continuous operation until a particular batch of paper had been finished. A stoppage had to be avoided if at all possible as Thomas Wrigley, an eminent Lancashire paper manufacturer explained nearly 40 years earlier.

> Whenever, from any cause, as it is, a paper machine has to be stopped, all the pulp must be washed off the wire, and the wet sheet scraped off the roller, and the sheet upon the cylinder breaks; then in starting afresh the pulp has to be led by the hand on to the wet felt, and thence again to the dry felt and through the presses. Again, one uniform heat can be easily maintained, but it is difficult to be got at once: if the cylinder is too hot, the paper snaps, if too cold, it is damp. We have 11 paper machines, and each machine has from six to eight cylinders; in other places each may have 10 or 12 and more; all the cylinders must be at exactly the same temperature before a perfect sheet can be made.[36]

The skilled and responsible tasks performed by beatermen and machinemen were reflected in their earnings and they were the best paid manual workers in the paper mill below the rank of foremen. In Scottish mills in 1906, when comprehensive national data is available for the first time, foremen averaged 45s 10d per week, with the earnings of the best paid 25 per cent exceeding 60 shillings a week. The average earnings of beatermen who received a bonus on top of their standard hourly rates of pay when production exceeded a certain figure were 35s 7d per week with the best paid 25 per cent earning more than 44s 6d per week. The comparable figures for machinemen where a bonus scheme was operating were an average of 37s 2d per week and, like the beatermen, 44s 6d for the best paid.

However, nearly half the beatermen and a quarter of the machinemen recorded in the Board of Trade survey were paid entirely on time rates and the absence of a bonus reduced their weekly average earnings to 30s 8d and 32s 2d respectively.[37]

Engineering workers, the other group of highly skilled male workers in the paper mill, had average earnings of 35s 8d per week, which was appreciably higher than beatermen and machinemen paid simply by the hour, but no better than those who were on a bonus system. Moreover, the best paid mechanics, the top 25 per cent earning more than 37 shillings a week, could not match the level of more than 44s 6d achieved by 25 per cent of both the beatermen and machinemen whose pay was boosted by bonuses.

At the other end of the scale unskilled male labourers aged 20 years and above working full time in Scottish paper mills had average earnings of only 19s 3d per week. The poorest 25 per cent earned less than 18 shillings, whilst the top 25 per cent exceeded 20 shillings a week. The relatively low paid male workers also included washers, breakers and bleachers whose average earnings were 21s 10d a week, calenderers and glazers with average earnings of 22s 6d, and boilers, warehousemen and packers whose earnings were little more than 23 shillings a week.

Overall, when allowance is made for the numbers in the different occupations, the average earnings of an adult male worker working full time in paper mills in Scotland in 1906 amounted to 25s 3d per week. He was thus less well paid than his counterpart in paper mills in the North of England or the United Kingdom as a whole where the comparable figure was 29 shillings a week. His earnings were also below the level of the typical manual worker in other industries in the United Kingdom that year where the average adult male earnings for full time work amounted to 30s 6d.[38]

The best paid 25 per cent of beatermen and machinemen who were on a bonus system approached the earnings of the top 10 per cent of male manual workers in the United Kingdom who earned more than 46 shillings a week. On the other hand labourers in Scottish paper mills ranked among the bottom 10 per cent of UK manual workers earning less than 19s 6d per week.

As in other industries women's earnings were substantially lower than those of men. The average earnings of all female manual workers aged 18 years and above in Scottish paper mills were no more than 12 shillings a week. Paper sorters in the finishing department on piece rates were the best paid. Their average earnings were 14s 10d per week, the top 25 per cent of those recorded exceeded 16s 6d and 6 per cent earned 20 shillings or more, which brought their earnings above the level of the poorest paid adult male workers, the unskilled labourers in the mill. Paper sorters paid hourly rates were less well paid, averaging 11s 3d per week, whilst at the bottom end of the scale, rag sorters averaged 10s 2d and rag cutters 9s 11d per week.

The earnings of the younger male workers, those under 20 years of age, if we exclude lads and boys employed as paper cutters who averaged only 8s 3d per week, were comparable with those of most women workers. However the earnings of girls under 18 years of age were significantly lower. As a group their average weekly earnings were 7s 2d and if piece workers are excluded the figure for the substantial majority who were paid hourly rates falls to 6s 4d whilst the poorest paid 25 per cent earned no more than 5 shillings a week.

The proportion of low paid workers at Mugiemoss was increased by the substantial number of female workers employed in bag making, an activity which was missing from the average paper mill. Rates of pay for the separate occupational groups in the papermaking department at Mugiemoss are not known at this period but the annual wage bill for bag makers at the mill in 1908 was estimated to be £7,200, which was equivalent to annual earnings of £27.27 per employee or approximately 10s 6d per week. This ranked them with rag sorters and rag cutters near the bottom end of the scale for female workers.[39]

ASPECTS OF WORK AND WELFARE

Other aspects of working life, though less dominant than the level of pay, were not unimportant. Hours of work and their distribution, holidays, the working environment, and the proximity of the workplace also require examination.

Many of the workers at Mugiemoss lived in the immediate neighbourhood and the Census Enumerators for the parish of Newhills at 10 year intervals in the second half of the nineteenth century recorded many scores of paper workers living within a mile and a half radius of the mill. The population of the parish increased from 2,255 persons in 1841 to 5,526 in 1891, providing a growing reservoir from which the mill could draw its workforce, albeit one also tapped by other local industries and probably also by the paper mills of Alexander Pirie & Sons of Stoneywood, located scarcely a mile away.[40]

Some of the Davidson workers lived in rented houses owned or held on long lease by the firm. A number of dwelling houses came into the company's possession as part of the lease of Bucksburn mills in 1876, some freehold property in Greenburn Road valued at £557 was acquired ten years later, and in 1917 there was further investment in houses in the same road.[41]

The houses in Greenburn Road and Mugiemoss Road were only a short distance from the mill but the housing stock was never sufficient to provide accommodation for more than a small proportion of the labour force and it would appear that the directors viewed their modest investment as a means of ensuring that key workers lived near their work and perhaps as an extra attraction when recruiting labour. Thus William Don of Newbattle Mill, Dalkeith, in October 1880 was offered the post of manager at a salary of £156 a year plus a free house and in May 1884 C. Kitchin was engaged as a foreman mechanic at 50 shillings per week and a free house after another applicant the previous month had been offered the same salary but with no house. Kitchin was informed that 'as one is not vacant at the moment provide yourself with one at our expense as far as rent goes, and we will give you the tenancy of the first vacant house that we have suitable'.[42]

In the years of full employment after the Second World War when skilled labour was in scarce supply, the directors took a further look at the housing supply. Shortages of bricks and imported timber meant that new building was controlled by Government regulations but the company obtained licences to erect six houses in 1950 – four 'ordinary type' at a cost of £1,800 each and two of the 'managerial type' at a cost of £2,100 each.[43]

The expansion of public transport services in Aberdeen and its suburbs in the final quarter of the 19th century widened the potential area from which Mugiemoss could draw its workforce, although the cost of travel even with special cheap fares deterred many workers from commuting daily to work. The tramway network of horse-drawn trams was extended to Woodside, approximately two miles from the mill, in 1880. Aberdeen Corporation, which bought out the private company running electric trams in August 1898, had a service operating to Woodside at the end of the following year.[44] The Aberdeen Suburban Tramways Company, formed in 1902, had completed its tramways to Stoneywood Church in Bankhead two years later and by agreement with Aberdeen Corporation, who supplied the electric power, it was running trams from St Nicholas Street in the centre of Aberdeen to Woodside and thence along Inverurie Road and Old Meldrum Road to the Bankhead Terminus, passing Mugiemoss en route. The service included tramcars running before 7.00 a.m. and after 5.00 p.m. at a special fare of ½d per mile for the benefit of 'the labouring classes'.[45]

Meanwhile the Great North of Scotland Railway Company, whose station at Bucksburn was only half a mile from the mill, had introduced a regular suburban service. By the beginning of 1888, after opening new stations at Don Street, Bankhead and Stoneywood the previous year, there were 12 trains a day in each direction running the six and a quarter miles between Aberdeen and Dyce, stopping at each station and taking 25 minutes for the journey. Bradshaw's Railway Guide for July 1887 lists the first train of the day leaving Aberdeen at 5.25 and arriving at Bucksburn at 5.40 and at Stoneywood at 5.46 in time for the start of the working day at 6.00 a.m.[46]

Of the real importance of all these services to the labour force at Mugiemoss, we can only surmise. However it is significant that when a journalist from the *Paper-Maker* made an extensive tour of Alexander Pirie and Sons' Stoneywood works, which was about a mile further from the city, in 1901 to prepare an article on that firm, he commented on the kitchen and dining room provided for the use of the workforce, adding that many of the workers resided in Aberdeen.[47]

Hours of work in the paper industry as in industry and commerce generally were excessively long by modern standards. The typical working week for full time workers in the United Kingdom in the middle of the 19th century was between 60 and 72 hours a week.[48]

In 1833 the first effective Factory Act passed by Parliament limited the hours of young persons working in textile factories except silk, to 8 hours per day for children between 9 and 13 and to 12 hours per day for those between 14 and 18, forbidding entirely employment under the age of 9. Night work was prohibited under the age of 18. Later legislation reduced these hours and then in 1867, when paper mills were included, and 1878 extended the scope of the Acts beyond textile factories to cover factories and larger workshops generally. Meanwhile the restrictions on the employment of young persons were re-inforced by other legislation which ensured elementary education nationwide with compulsory attendance from 1880 to the age of 10, raised in 1893 to 11 and to 12 at the end of the century.[49]

Night work for women was also forbidden under the Factory Acts and problems of enforcement led Parliament in 1850 and 1853 to define the normal working day. Women and young persons were to work only between 6 a.m. and 6 p.m. or between 7 a.m. and 7 p.m. with one and a half hours for meals and were to cease work at 2 p.m. on Saturday.[50]

The Factory Acts did not limit the hours of adult males although the gains made by women and young persons undoubtedly encouraged their own efforts. Over the 60 years between the middle of the 19th century and the eve of the First World war the length of the working day for adult workers as a whole was reduced substantially and by 1914 the average working week of full time workers had fallen to some 56 hours.[51]

The hours worked in the paper industry in the early years of this century were greater than the national average. In the United Kingdom as a whole the average for more than 15,000 workers employed full time who were surveyed in the comprehensive study made by the Board of Trade in 1906 was 56.9 hours a week, exclusive of meal times and overtime, with workers in the North of England averaging 58.2 hours and those in Scotland 57.7 hours a week. Over 73 per cent of the Scottish paper workers had hours very close to this average, ranging between 56 and 58 hours, whilst 16 per cent worked for 60 or more hours a week.[52] Both shift workers and persons working regularly in the daytime were included in the survey. The latter comprised workers in the preparatory processes and in the finishing departments. For the shift workers, the pattern of work was governed by the continuous operations of the paper-making machine. Machinemen and beatermen and their assistants with a small support force which included stokers manning the steam boilers and some labourers, until the end of the First World war worked 12 hour shifts, alternating between night shifts in one week and day shifts in the next. Unlike the permanent day workers they had no definite meal breaks and food had to be eaten, often with great haste, when the opportunity arose.

Throughout the nineteenth century until the 1890s the day shift typically began at 6 a.m. each day Monday to Saturday but the last shift of the week might be extended, depending on the wishes of the particular mill owner, to 8 p.m. or even 10 p.m. The night shift ran from 6 p.m. Monday to Friday, the week commencing with a half shift from midnight on Sunday to 6 a.m. on Monday. The shift workers thus worked as much as 78 hours one week and 66 at night the next, giving a total of 144 hours per fortnight.[53]

However, the practice of half day working on Saturday spread steadily in British factories during the second half of the century and soon became widespread, making Britain the envy of European workers. In the 1890s some paper mill owners began to adopt the practice, closing their mills at 2 p.m. on Saturday, and by 1906 the Board of Trade enquirers assumed that the normal hours for shift workers were 132 per fortnight, consisting of 72 hours in day work and 60 hours at night. Nevertheless a national agreement on Saturday closing drawn up by the employers and the National Union of Paper Mill Workers was not signed until 1913. This stipulated that day workers would cease work at 12 noon, that paper-making machines would stop 1 p.m. and that wires and felts would be put on during the week so that only in exceptional circumstances would it be necessary for work to be done after 2 p.m. on Saturdays.[54]

Further improvements followed after the First World War. In 1919 a national agreement with the employers introduced 8 hour shifts worked in a three week rotation making a total of 132 hours or 44 hours a week, and laid down a working week of 48 hours for permanent day workers. But no other reductions in hours, holidays excepted, were negotiated nationally until 1953.[55]

Holidays made a very marginal addition to total leisure time. Even in 1906 the average number of days holiday per year for paper workers in Scotland was less than seven, the actual number normally ranging from six to eight. In Yorkshire, Lancashire and Cheshire up to eleven days a year might be taken in holidays, with eight days on average. Holidays with pay – limited to six days in the year – were not agreed nationally until shortly before the outbreak of the Second World War.[56]

Specific information about working hours at Mugiemoss is extremely sparse. Hours for shift workers in the paper industry in the Aberdeen area in 1883 were 72 per week, as compared with 57 for day workers, and this was equivalent to the fortnightly cycle of

144 hours, comprising 78 hours by day one week and 66 of night work the following week, then typical in the industry. William Ross, General Secretary of the National Union of Paper Mill Workers, giving evidence to the Labour Commission in 1891 considered that these hours, which still prevailed in the North of Scotland although most paper mill employers in Scotland had recently reduced their shift workers' hours to 132 per fortnight, 'were most excessive'. However the hours of the permanent day workers in 1883 seem to have been similar to those worked in other districts in Scotland and were in line with the figures for the whole of Scotland noted in the later survey of 1906.[57]

In the 1930s, although the firm did not recognise the paper workers' trade union and refused to give it negotiating rights, an examination of the first and only surviving Wages Book from Mugiemoss, covering the two years from February 1930 to June 1932, reveals that the hours of the permanent day workers did not exceed the figure of 48 per week laid down in the national agreement of 1919. Shift workers on a few occasions when the mill was busy worked longer hours than the nationally agreed figure, some individuals averaging 47 or even more hours a week over the three week cycle, apparently without overtime payments. However there is no indication that the management were still operating 12 hour shifts which was one of the points in dispute four years later in the strike of 1936 and underlines Colonel Davidson's claim that a 12 hour shift had been employed as a temporary measure only when the new Board Machine was being introduced.[58]

Working conditions within the paper mill varied considerably. In most manual jobs considerable physical effort was required throughout the nineteenth century. Labourers trundled barrows full of various materials, china clay and other additives were added to the beater manually, and in the finishing department women sorted piles of paper which they had lifted onto the surface of the benches at which they stood. Perhaps the most unpleasant conditions were to be found in the rag sorting and cutting room where clouds of dust filled the air, threatening damage to the lungs if prolonged over a long period, and at the wet end of the papermaking machine where noise, heat, and steam combined to create an oppressive atmosphere.

There was also the risk, as in most factories, of industrial accidents. Workers at opposite ends of the production process cutting rags or finished paper sometimes sliced through their fingers. More serious was the danger posed by the continuous operation of the power driven paper-making machine which was being driven at speeds of up to 80 feet a minute by the early 1860s and at speeds three times as great forty years later. It was possible to fence off parts of the machine so that no-one accidentally fell or leant into it, but this did nothing to remove one of the common causes of accidents which occurred when the machine was re-started after a breakage since one of the operatives had to lean over the machine and lead the wet sheet of pulped fibre by hand from the wire belt onto the wet felt then to the dry felt and over the drying cylinders.[59]

The extension of the Factory Acts to include paper mills and some other factories not making textiles in 1867 marked the beginning of official pressure to improve ventilation, control excessive dust, and fence dangerous machinery. William Ross, giving evidence to the Royal Commission on Labour in 1892, complained about poor ventilation in the paper-making machine rooms and asserted that more action could be taken to avoid fatal accidents in the industry, adding that he had no great faith in the factory inspectors. However under cross examination he admitted that his trade union had not made any complaints about specific cases to the inspectors and he conceded that 'they are always at hand and you can always communicate with them without even giving names'.[60]

In the Aberdeen area a few years later, there was certainly no evidence to suggest that the factory inspectors were inactive. In 1899, for example, Culter Mills Paper Co. Ltd. and Donside Paper Co. Ltd. were successfully prosecuted for failure to fence some dangerous machinery whilst Mary M. Paterson one of the relatively new breed of peripatetic female inspectors, who regularly travelled more than 10,000 miles a year throughout the United Kingdom in pursuit of her duties, made an inspection at Mugiemoss soon after this. She subsequently wrote to complain of 'the excessive dust in the rag-sorting, dusting and cutting departments in which you employ a large number of women' and she requested that the firm consult ventilating engineers in order to instal a fan or other mechanical means for removing the dust. In reply the company secretary regretted that she had only spoken to the foreman on her visit and claimed virtuously that 'we have always found that instead of the dust arising from the process having any injurious effect on the girls, they have always been the strongest and healthiest of all the girls employed at these works'. He then added 'We have not therefore thought it necessary to consider the matter of a fan, especially as we believe that the Fire Insurance Companies would object to anything in the form of a fan, which would create a strong draught, because in this department there is always a danger of the rag cutters striking on metal or other hard substances which may be concealed amongst the rags, and the spark thus caused would be a source of great danger if there was a strong current of air to fan it, or suck it upwards'.[61]

The concern shown by various observers over safety and health in the paper industry however should be set against a wider perspective by examining mortality statistics for industries and occupations generally in the United Kingdom. Dr. Hunt, surveying the data on comparative mortality among occupied males from 1900 to 1902 published by the Registrar General, pointed out that men working in the paper industry were decidedly more healthy than average, with a record akin to that of teachers and farm labourers and quite different from such occupations as tin miners, general labourers, file makers, seamen, inn keepers and publicans or cutlers and scissor makers at the very opposite end of the scale.[62]

A.D. Spicer who wrote the first lengthy scholarly history of the paper industry in 1906 was in no doubt that conditions within the industry had greatly improved during the nineteenth century. He cited the provision of suitable dining rooms and lavatories, the installation of efficient dust extractors in the rag dusting and cleaning operations, the provision of steam pipes to warm the paper sorting departments in cold weather, and the 'adoption of travelling cranes and similar appliances' to reduce the need for heavy labour. He also observed that the modern mill buildings were lofty and well ventilated, employing exhaust fans to remove the steam from the premises which housed the paper-making machine.[63]

Appendix A

The Directors of C. Davidson & Sons Ltd. 1875–1953

	Name	Years in Office	Position held when he joined the Board [1]
1	Davidson, Alexander (1842–1920)	1875–1920	Son of William Davidson (GSF)
2	Davidson, Alan John (1876–1966)	1921–1926	Son of Alexander Davidson
3	Davidson, Charles William (1869–1943)	1898–1943	Son of Charles Davidson Jnr. (Son of William Davidson)
4	Davidson, David (1848–1915)	1875–1915	Son of William Davidson (GSF)
5	Davidson, David Peter [2] (1910–1986)	1946–1953	Son of Thomas Davidson
6	Davidson, John (1845–1897)	1875–1897	Son of William Davidson (GSF)
7	Davidson, George (1838–1875)	1875	Son of William Davidson (GSF)
8	Davidson, Thomas (1880–1951)	1911–1951	Son of David Davidson
9	Davidson, William Dalzell (1871–1940)	1898–1926	Son of Alexander Davidson
10	Dawson, Alexander Thomson (1885–1968)	1926–1946	Director, Morrison's Economic Stores Ltd
11	Dawson, James Mearns (1914–)	1946–1953	Son of A.T Dawson
12	Duffus, James Catto [2] (1891–1962)	1936–1953	Partner, Wilsone and Duffus
13	Ledingham, Robert Mackay [2] (1893–1969)	1951–1953	Partner, Edmonds and Ledingham

14 Mackie, John (1849–1911)	1898–1911	Manager, London Office
15 Marr, Alexander (1845–1930)	1908–1926	Director, Alexander Pirie and Sons Ltd.
16 McCrone, Robert Watson (1893–1982)	1936–1946	Managing Director, Metal Industries Ltd. Director, British Oxygen Ltd.
17 Mellis, William Ranald[2] Stewart (1909–1970)	1946–1953	Managing Director, Mitchell & Muil Ltd.
18 Montgomerie, John Alexander (1877–1957)	1936–1946	Managing Director, Montgomerie, Stobo & Co Ltd, Director, International Bitumen Emulsions Ltd
19 Partington, James (1911–1981)	1948–1953	Company Secretary
20 Spence, Henry Reginald[2] (1897–1981)	1951–1953	Managing Director, William Spence and Son, Managing Director, Harrott & Co Ltd.
21 Tawse, Bertram William[2] (1906–1980)	1951–1953	Chairman, William Tawse Ltd.
22 Warburton, Eric Jackson (1905–)	1951–1952	Mill General Manager
23 Williamson, Frank (1898–)	1943–1950	General Manager

(1) GSF = Grandson of Founder, Charles Davidson
(2) Appointed to the new board of Directors after the merger in 1953.

Appendix B

The Dawson Shareholdings 1925–1938

A.T. Dawson purchased his first shares in C. Davidson & Sons Ltd. in January 1923. In 1925 the annual return of shareholders recorded a holding of 150 shares whilst his wife, Isobel, held 100 and his brother Rupert G. Dawson, a further 750. Over the next nine years the number of shares held by his wife and his brother remained unchanged but he steadily increased his holdings to 2,030 shares in 1927, rising to 4,295 in 1929 and 9,005 shares by December 1934.[1]

The relative shareholdings of the Dawson and Davidson families in December 1934, the year before the financial crisis and drastic capital reconstruction of the company are summarised in Table 30.

The reduction of share capital in 1935, which wrote down the nominal value of each £1 share to one shilling, left the relative share of the Dawson family in the equity capital unchanged at 10 per cent. However, the issue of an additional 360,000 shilling shares in 1937, which increased the equity by £18,000, completely transformed the position. In January 1938 the total Davidson shareholdings were virtually the same as in 1934, amounting to 21,834 shares. On the other hand the Dawson family had made substantial purchases of the extra shares, increasing their overall holdings to 310,611 shares, which now represented almost 68 per cent of the issued share capital, including more than 44 per cent held by A.T. Dawson alone. Their holdings are summarised in Table 31.

APPENDIX B NOTE

1 Share Ledger 1923–1935, Annual Return of Shareholders 1924–45

Table 30: The Dawson and Davidson Shareholdings, December 1934

Name of Shareholders	*Number of £ Shares Held*
1 The Dawson family (A.T. Dawson, his wife and his brother)	9,855
2 Colonel Thomas Davidson and family (T. Davidson, his wife and his son)	5,304
3 Charles William Davidson and his wife	2,438
Combined holdings of the 3 Directors and their families	17,597
4 Former Davidson Directors and their families[1]	
W.D. Davidson and his wife	3,126
A.J. Davidson	912
5 Other Davidson Holdings[2]	
Sir N.G. Davidson, brother of A.J.D.	1,012
Joint holdings as executors of various wills	9,032
	31,679
6 Total Number of Shares issued	97,926

Source: Annual Return of Shareholders, 1924–1945

Notes: 1 W.D. Davidson was living in Aberdeen, A.J. Davidson in Putney, London, SW15

2 Sir N.G. Davidson was living in Clapham Village, Sussex.

Table 31: The Dawson shareholdings, January 1938

Name of Shareholders	*Number of one shilling shares held*
1 A.T. Dawson	204,761
2 His wife, Isobel	100
3 His father, Peter[1]	40,000
4 His father's wife, Jeannie	20,000
5 His sister, Margaret Smith Dawson	40,000
6 His brother, Rupert G. Dawson[2]	5,750
Total	310,611
7 Total number of Shares issued	457,900

Source: See Table 30

Notes: 1 His father, wife Jeannie, and sister were living at Drum Coille, Braco, Perthshire.

2 His brother was living at Orchil, Braco, Perthshire.

Appendix C

The 1936 Strike

The rescue operation which saved the firm from bankruptcy in 1935 and raised fresh capital to develop the production of millboard and of Ibeco paper overcame various problems before success was achieved. However one obstacle was quite unexpected, a strike at Mugiemoss in October 1936 which lasted for three months. This was the only major labour dispute to have come to light in the history of the firm and for this reason alone it deserves some examination.

The seeds of the conflict had been sown ten years earlier when the National Union of Printing, Bookbinding, Machine Ruling and Paper Workers decided to support the T.U.C. action in support of the coal miners and took part in the General Strike in May 1926. The other trade union embracing paper-workers, the Amalgamated Society of Paper Makers, although not affiliated to the T.U.C., resolved that its members in mills where the National Union had acted would fall into line. The strike was called off after 9 days and subsequently the general secretary of the Amalgamated Society reported to his executive that 'Generally speaking the employers in the paper trade have acted without bitterness, and have reinstated their workers, which everyone must appreciate.'[1]

However the strike had caused considerable disruption to output in the paper mills and the Scottish papermakers meeting in Edinburgh on 12th May, the day the strike ended, resolved to have no further dealings with the two unions. The management of the four Aberdeen paper manufacturing firms – Culter Paper Mills Ltd., C. Davidson & Sons Ltd., Donside Paper Co. Ltd., and Alexander Pirie & Sons Ltd. – met in Aberdeen the next day and confirmed this anti-union policy but agreed that the existing rates of wages and conditions would be maintained and that strikers might apply for re-engagement. The decision announced in the *Aberdeen Press and Journal* began by stating that the four firms 'Have resolved that, owing to the National Union of Printing and Paper Workers and Amalgamated Society having failed to keep their word by calling out their members without due notice, thus causing heavy loss to all mills, no future dealings will be held with these unions.'[2]

At Mugiemoss the end of the strike followed the personal intervention of Colonel Davidson, who addressed his workers standing on a stack of wood pulp in the yard. The official BPB history of 1973, drawing on the memories of men who were present, records that the mill resumed work the next day 'but on the understanding that no-one would carry a union card until the management was satisfied that the unions had put their house in order and that members in relatively small industrial centres, such as Aberdeenshire, would get a fair deal and not have to do what they were told by leaders in the South who knew nothing and cared less about local problems.[3]

The ban on the paper workers' unions remained in force at Mugiemoss over the following years. Nevertheless it would appear that relations between the workers and the management were relatively unharmed. Colonel Davidson genuinely seems to have cared for his workforce and workers of this period recalled in 1973 his honest straightforward manner and sympathetic understanding of their problems. Certainly the respect in which he was held helped to maintain morale and keep the firm alive in very difficult circumstances.[4]

It is ironic therefore that in 1936 just when the firm's debts had been reduced to an acceptable level, with fresh capital injected to develop new products which offered a secure future for Mugiemoss, a major strike should occur. However the measures taken to revive the firm's fortunes took place in an improving economic climate. Recovery from the depths of the world slump had begun in the United Kingdom late in 1932 and from then until the middle of 1937 there was a period of sustained growth in production and employment. Membership of both the paper workers' unions nationally was increasing again, and growing co-operation between them, which resulted in merger on 1st January 1937, strengthened their determination to improve the pay and conditions of paper mill workers throughout the United Kingdom.[5]

At Mugiemoss itself the appeal of union action was enhanced by the unions' growing strength elsewhere and perhaps also by a growing impatience among the workers to enjoy the better future which now seemed possible after years of restraint endured since the early 1920s. More specifically, the labour force had been swollen by labourers recruited for a few weeks only to assist with the installation of new plant for the production of millboard, whilst a few men directly involved with the new Board Machine were working 12 hour shifts in an attempt to overcome teething problems and to ensure that a satisfactory level of output was achieved as soon as possible.[6]

The Donside Branch of the National Union of Printing, Book-Binding, and Paper Workers threw down the gauntlet on 26th September 1936 by giving the directors notice that unless full trade union recognition was granted a strike would commence a week later on 3rd October. They also demanded an 8 hour day and payment of proper rates of overtime. Other workers at the firm, including engineers, draughtsmen, painters and joiners, who belonged to unions which were recognised by the directors were not asked to come out in sympathy and took no part in the dispute.[7]

The strike began on 3rd October. Some 170 workers were involved and the Papermaking Department of the mill was paralysed by the stoppage. It was 12 weeks before the directors and strikers met on 31st December to discuss a settlement and it was not until 9th January 1937 that the men returned to work with the union still not recognised by the directors.

One of the strikers bitterly told the local Press that 'Our position was greatly weakened by the action of some of the men who filtered back into the mill. . . . Every week our ranks were depleted and at the same time new hands were being engaged'.[8] However it is apparent from the Daily Report Book covering this period, which is summarised in Table 32, that the strike was very effective initially and well supported for at least two months. Even in the best week of October, the three paper-making machines were operating for no more than 5 shifts in total compared with a possible maximum of 51, and during the whole month only 13 shifts were worked, yielding a derisory output of 14 tons of paper, whilst the Board Machine was completely inactive. It was not until the second week in December that the paper-making machines were regularly operating a total of 20 or more shifts per week. The Board Machine

Table 32: The Effect of the 1936 Strike on Paper and Millboard Production

	Machines 1–3		Machine No.4	
Week Ending Saturday	No. of Shifts Worked	Output of Paper (Tons)	No. of Shifts Worked	Output of Millboard (Tons)
10 October	5	3	—	—
17 October	—	—	—	—
24 October	3	4	—	—
31 October	5	7	—	—
7 November	6	14	—	—
14 November	10	23	—	—
21 November	12	30	3	21
28 November	12	25	6	31
5 December	16	27	6	35
12 December	20	36	6	53
19 December	22	64	6	42
26 December	22	28	3	62
2 January	14	40	4	27
9 January	26	68	7	41
16 January	48	83	8	65
23 January	51	111	11	44

Source: Daily Report Books 1936–1937

was at work from 19th November but it operated for only a single shift per day for the remainder of the year.

Towards the end of the strike, on 6th January, Colonel Davidson addressed a mass meeting of the strikers at the mill but a letter sent by the firm to individual strikers was perhaps more persuasive. They were informed that if they wished to return to their former jobs they must apply by mid-day on Saturday 9th January. After this deadline the management 'will proceed to fill up its vacancies at its own convenience and will accept applications from any source'.

The union put a brave face on the outcome. W.T. Airlie, the union organiser, claimed that improved rates of pay had been secured and also adherence to an 8 hour day, although as was noted in Section 4 of Chapter 8, this had in essence been achieved at Mugiemoss some years earlier. He added that a new system of sanitation would soon be installed at the mill, implying that this was another of the gains achieved by the strike.

Colonel Davidson assured the *Aberdeen Press and Journal* that there would be no victimisation and promised that 'he would take back all the fellows whose jobs are still open'. In fact approximately 50 of the strikers were taken on at the mill when they applied for re-employment whilst the names of others were placed on a list at the mill lodge in the expectation that work would become available for some of them shortly, although a considerable proportion of the 120 strikers not re-engaged immediately had been working as labourers on a temporary basis whilst new plant was being installed.[9]

However the real hope of healing the wounds caused by the dispute lay in the future. Rising sales as the new products were developed would generate additional employment and Colonel Davidson was especially optimistic about the new Board Machine, envisaging that once the mill was fully at work again, the total labour force would increase from some 350 persons employed before the strike to a total of some 410–420.[10]

APPENDIX C NOTES

1 Bundock, pp 329, 335, 426
2 *World's Paper Trade Review*, Jan–June 1926, p. 1588, *Aberdeen Press and Journal* 14th May 1926
3 *The History of BPB Industries*, p. 86
4 *Ibid.*, p. 85
5 Aldcroft pp 43–4, Bundock pp 353–4, 435, 442–3
6 *Aberdeen Press and Journal*, 5 October 1926, 9 January 1937. See also AUA, MS 2649, 1 & 2
7 *Aberdeen Press and Journal*, 5,6 October 1926
8 *Ibid.*, 1,9 January 1937
9 *Ibid.*, 7,8,9,11 January 1937
10 *Ibid.*, 8 January 1937

Notes

The following abbreviations have been used.

AUA	Aberdeen University Archives
BC	Birth Certificate
CRO Edinburgh	Companies Registration Office, Edinburgh
DA	Archives of C Davidson & Sons Ltd
DC	Death Certificate
HC	House of Commons
MC	Marriage Certificate
PP	Parliamentary Papers
PRO	Public Record Office
SRO	Scottish Record Office

Chapter 1 The Early Days

1 A.G. Thomson, *The Paper Industry in Scotland*, pp. 74–9, 118–19.
2 A. Keith, *A Thousand Years of Aberdeen*, pp. 308–10, 358.
3 D.C. Coleman, *The British Paper Industry, 1495–1860*, pp. 109–12, Thomson, pp. 41–3.
4 *World's Paper Trade Review*, 29 Aug. 1890, 6 June 1913, Thomson pp. 118, 148–49.
5 P. Morgan, *Annals of Woodside and Newhills*, p. 187.
6 S.R.O., Sheriff Court Books for Aberdeenshire, 7 June 1811.
7 S.R.O., Sheriff Court Books for Aberdeenshire, 22 April 1826 and 15 Nov. 1833.
8 Morgan p. 187, Thomson, *frontispiece* and p. 177.
9 S.R.O., Sheriff Court Book for Aberdeenshire, Inventory of Charles Davidson 18 Oct. 1844.

10 *New Statistical Account of Scotland*, XII, 72, 239.
11 Morgan, p. 187 *World's Paper Trade Review* 18 April 1913.
12 DA, Private Ledger 1852–1865, p. 1.
13 *Ibid.*, p. 6, *Aberdeen Journal*, 6 July 1853.
14 AUA, MS 2769, I/30/5, Memorial for . . . opinion of Counsel 3 March 1875, Accounts for Half Year ending 31 July 1858.
15 DA, Private Ledger 1852–1865, pp. 7, 8, 16, 20, 26.
16 Morgan p. 188, Thomson pp. 207, 209.
17 See for example B. Darwin, *Robinsons of Bristol 1844–1944*, pp. 18, 19, 25, *The Grocer* 27 June 1863, 6 Jan. 1877, advertisements by C.T. Youngman, James Baldwin & Sons, *World's Paper Trade Review* Jan–June 1890, p. 513, William White, *History, Gazetteer and Directory of Suffolk*, 1874 pp. 28, 37.
18 *British Patents* 1859 No. 2724, 1863 No. 1476.
19 DA, Private Ledger 1852–1865, pp. 49, 116–17, 130–31.
20 DA, Inventory of Title Deeds, 1877, lease of 80 Upper Thames Street 1858, Private Ledger 1852–1865 p. 9.
21 DA, Private Ledger 1852–1865, pp. 35–6, 42–3, 118.
22 Coleman p. 331.
23 DA, Private Ledger 1852–1865, pp. 35–6.
24 AUA, MS 2769, I/30/5, Accounts for Half Year ending 31 July 1858.
25 AUA, MS 2769, I/30/1, Minute Book 1871–1875, I/30/5, Half Yearly Accounts 1865–1875.
26 AUA, MS 2769, I/30/5, Half Yearly Accounts ending 31 July 1870, 31 Jan. 1872, I/30/1, Minute Book 1871–1875 pp. 9, 11, 15, 21, 22, 26, 40.
27 AUA, MS 2769, I/30/5, Half Yearly Accounts ending 31 Jan. 1870, I/30/1, Minute Book 1871–1875 pp. 10, 15, 25.
28 AUA, MS 2769, I/30/1, Minute Book 1871–1875 pp. 14, 20, 21, 25, 26, 33, 34.
29 AUA, MS 2769, I/30/5, Half Yearly Accounts ending 31 Jan 1868, 31 Jan, 1875, I/30/1, Minute Book 1871–1875, p. 43.
30 S.R.O., Sheriff Court Books for Aberdeenshire, Inventory of William Davidson, 24 Nov. 1873.
31 DA, Private ledger 1852–1865, p. 2, AUA, MS 2769, I/30/5, Half Yearly Accounts ending 31 Jan. 1873.
32 *Fourth Report of the Commissioners appointed in 1868 to inquire into the best means of preventing the pollution of rivers. Pollution of Rivers*

in Scotland. PP 1872, XXXIV, Evidence, Answers to Queries Part ii, 225.

33 *Paper Mills Directory*, 1871.

34 Morgan, pp. 187–188, *World's Paper Trade Review*, 18 April 1913, *Aberdeen Journal*, 7 Jan. 1896.

35 AUA, MS 2769, I/30/5, Memorial for . . . opinion of Counsel 3 March 1873, pp. 1–3.

36 *Ibid.* pp. 3–6.

37 *Ibid.* p. 7.

38 *Ibid.* pp. 7–28 and opinion of Counsel attached.

39 P.L. Payne, *The Early Scottish Limited Companies, 1856–1895,* pp. 19, 31–2.

40 DA, Memorandum on the proposed conversion of Charles Davidson & Sons into a limited company.

41 DA, Memorandum and Articles of Association, 1875.

Chapter 2 Expansion and Prosperity, 1875–1900

1 DA, Inventory of Title Deeds, 1877: lease of Bucksburn Mills, Minute of resolutions passed at Extra-ordinary General Meeting 27 Jan. 1876, *Annual Report and Accounts*, 1876, S.R.O. Register of Sasines, Aberdeenshire 2 Feb. 1876.

2 AUA, MS 2769 1/30/1, Minute Book 1871–1875, pp. 14–15, 21, DA, Letter Book of John Davidson, 1 Nov. 1872, *Annual Report and Accounts*, 1876, Morgan pp. 188–9.

3 DA, Inventory of Title Deeds, 1877, leases of 51 and 59–63 St. Enoch Square, Glasgow.

4 *Ibid.* leases of 34 Dean Street and office on High Bridge, Newcastle, *Paper Trade Review* 13 Aug. 1886.

5 DA, *Annual Report and Accounts*, 1888–1894, Inventory of Title Deeds, 1877, lease of warehouse in Castle Terrace, Edinburgh.

6 DA, Inventory of Title Deeds, 1877, lease of 80 Upper Thames St., London, Inventory of Title Deeds, 1914, lease of Paul's Pier Wharf, London.

7 AUA, MS 2769, I/30/1, Minute Book 1871–1875, p. 16, *Aberdeen Journal* 7 Jan. 1896, DA, Letter from Robert S. Cumming 23 Feb. 1898 and reply by the company secretary,

Director's Minute Book 1898–1900, monthly summaries of sales.

8 *Paper Mills Directory*, 1886, DA, Agreement for the sale of the Boulinikon Felt Co. Ltd., 27 Sept. 1904.

9 AUA, MS 2769, I/30/1, Half Yearly Accounts ending 31 Jan., 31 July 1874, 31 Jan. 1875.

10 PRO, Board of Trade, Files of dissolved companies, The Colonial Paper Co. Ltd., *Memorandum and Articles of Association*, DA, Agreement for the sale of Davidson's business in New South Wales, 8 Nov. 1893.

11 *Paper Mills Directory*, 1876, 1890.

12 Information about the characteristics and uses of various kinds of paper at this period has been derived mainly from E.A. Dawe, *Paper and its Uses*, and R.W. Sindall, *Paper Technology: an elementary manual.*

13 Dawe, pp. 30–2, Sindall pp. 8, 25, 43–4.

14 Dawe, pp. 116, 134, DA, Letter to Liverpool branch 1 Aug. 1913 and correspondence with Mitchells, Ashworth, Stansfield & Co. Ltd., Manchester, in November 1913.

15 DA, Memorandum and Articles of Association of the Boulinikon Felt Co. Ltd., 1883.

16 *British Patent*, 1884 No. 4897.

17 DA, Bundle of letters of January 1895 labelled 'Painted Roofing Felt' and circular from H.C. Petersen & Co., Copenhagen.

18 *British Patents* 1884 No. 593, 1898 No. 2327, *Aberdeen Journal* 7 Jan. 1896, *World's Paper Trade Review* Jan.–June 1898, p. 505.

19 DA, Agreement between the Boulinikon Felt Co. Ltd. and Mitchells, Ashworth, Stansfield & Co. Ltd., 27 Sept. 1904.

20 DA, Draft Agreement with Job Duerden & Co. 21 May 1887, and associated correspondence.

21 *World's Paper Trade Review*, July–Dec. 1893, pp. 479–80, CRO Edinburgh, Davidson Files, Agreement with Anglo American Self Opening Square Paper Bag Machine and Manufacturing Co. Ltd., 8 Dec. 1898, DA, Directors' Minute Book 1898–1900, 9 Dec. 1898, 9 Jan. 1899.

22 *British Patents*, 1893 No. 24092, 1895 No. 16418

23 DA, Directors' Minute Book 1898–1900, 18 Oct. 1898, Two lists of machinery in the Machine Bag Department, 1906–1924, Pamphlet produced to celebrate the firm's centenary, incorporating material from the *Aberdeen Journal*, 7 Jan. 1896.

24 *Ibid*, 2 July 1898, *Paper Mills Directory*, 1886, *Aberdeen Journal*, 7 Jan. 1896.

25 DA, Circular letter to Shareholders, 20 April 1891, *Annual Report and Accounts*, 1905, Draft Agreement with J.S. Bird 17 Sept. 1895 and associated correspondence.

26 *Paper Trade Review*, July–Dec. 1890, p. 390.

27 *Paper Mills Directory*, 1871, 1900, *Aberdeen Journal* 7 Jan. 1896, 'Mr Alexander Davidson, Mr John Davidson and Mr David Davidson', *Men of the Period, Scotland.*

28 J.N. Bartlett 'Investment for survival: Culter Mills Paper Company Ltd., 1865–1914', *Northern Scotland*, V, No. 1 pp. 48–50.

29 DA, *Annual Report and Accounts*, 1889–1893, 1899, 1900, Directors' Minute Book 1898–1900, monthly summaries of sales, *Paper Mills Directory*, 1895–1905.

30 DA, Annual Report and Accounts, 1876, 1882, 1887, 1888, CRO Edinburgh, Davidson Files, Resolution passed at Extraordinary General Meetings 6, 27 Feb. 1883. For a general discussion of the types and denominations of shares in industrial companies in the second half of the nineteenth century see J.B. Jefferys, *Business Organisation in Great Britain, 1856–1914*, chapters 4 and 5, and P.L. Cottrell *Industrial Finance 1830–1914*, pp. 80–8, 162–7.

31 DA, *Annual Report and Accounts*, 1877–1896, Register of Mortgages, bonds granted 14 Feb., 19 June 1876, 11 Aug. 1893. The Register records that the bondholders for the Bucksburn property were J.H. Harvey and J. Stephen, local farmers. The trustees of the late A.R. Dyer, an Aberdeen ship owner, provided a bond of £5,500 on the Mugiemoss property, whilst the Aberdeen legal firm of Milne & Walker, acting as trustees for various persons, provided another bond of £5,500 and one of £1,000.

32 DA, Circular letter to Shareholders, 20 April 1891, *Annual Report and Accounts*, 1891–1896.

33 Cottrell, pp. 164–6.

34 DA, Prospectus for issue of 4.5 per cent Debentures, May 1896, *Annual Report and Accounts*, 1897, Private Journal No. 2, 1895–1914, pp. 4, 7, 10. The total debenture issue expenses amounted to £881.

35 *Paper Mills Directory*, 1870–1900.

36 A.D. Spicer, *The Paper Trade*, pp. 13–24, 54–85.

37 *Annual Statement of Trade of the United Kingdom* P.P. 1873–1899,

C.F. Cross and E.J. Bevan, *A Notebook of Papermaking*, 2nd edition pp. 62–78, Sindall pp. 54, 68–9, J. Strachan 'The invention of wood pulp processes in Britain during the 19th century', *Paper-Maker*, Annual Number 1949.

38 *Paper Trade Review*, 28 Jan. 11 Feb. 1887, 6 Jan., 20 April, 11 May 1888, 8, 15 April, 13 May, 24 June 1898, 20 Jan. 1899, 11 May 1900, R. Evely and I.M.D. Little, *Concentration in British Industry*, p. 270.

39 Spicer pp. 89–90, 103–5.

40 *Paper Mills Directory*, 1880–1883, AUA, MS 2769, I/30/1, Minute book 1871–1875, p. 36, DA, Directors' Minute Book 1898–1900, monthly summaries of production. See also Chapter 1, Table 4.

41 *Aberdeen Journal*, 7 Jan. 1896.

42 DA, *Annual Report and Accounts*, 1876–1899.

43 *Paper Trade Review*, 17 June, 26 Aug., 9 Sept. 1887.

44 *Aberdeen Journal*, 20, 24 Nov. 1897, DA, *Circular letter to Shareholders*, 26 Feb. 1898, letters from George Bruce and Robert S. Cumming to W.E. Grassick, company secretary, in 1898.

45 DA, Memorandum and Articles of Association, amended by special resolutions passed at various dates Feb. 1883–Oct. 1918.

46 M. Edelstein 'Realised rates of return on U.K. Home and Overseas Portfolio Investment in the Age of High Imperialism' *Explorations in Economic History*, XIII (1976).

47 See my articles on these firms in *Northern Scotland* and *Business History*, listed in the Bibliography.

Chapter 3 Crisis and Recovery, 1900–1939

1 DA, *Annual Report and Accounts*, 1903–1908, *Memorandum and Articles of Association*, amended by special resolutions passed at various dates Feb. 1883–Oct. 1918, J.N. Bartlett, 'Alexander Pirie & Sons of Aberdeen and the Expansion of the British Paper Industry c. 1860–1914', *Business History* XXII, Number 1, p. 21. Marr joined the Pirie Board in 1904.

2 See Chapter 2, section 2. DA, *Annual Report and Accounts*, 1891–1912, Inventory of Title Deeds, 1914, Feu Charter of Mugiemoss and Bucksburn granted in 1911.

3 DA, *Annual Report and Accounts,* 1905–1909, *Paper Mills Directory,* 1908, *Directory of Papermakers of the U.K.*, 1910, 1911.

4 DA, *Circular letters to Shareholders* 9 March, 11 Sept. 1907, *Annual Report and Accounts,* 1907, Consumption of Coals, Mugiemoss, monthly summaries 1895–1923 (with annual paper production).

5 *World's Paper Trade Review,* 1 Jan. 1909, 18 April 1913, *Directory of Papermakers of the U.K.*, 1910. For a general description of an MG Papermaking Machine see Dawe, p. 25, Sindall p. 81.

6 DA, *Annual Report and Accounts,* 1909–1913, Private Journal No. 2, 1895–1914, pp. 61–2, Invoices for M.G. Machine from Bertram's Ltd.

7 DA, *Annual Report and Accounts,* 1905.

8 DA, *Circular letters to Shareholders,* 11 Sept. 1907.

9 *World's Paper Trade Review,* 18 April 1913, *Directory of Papermakers of the U.K.*, 1913, Dawe, pp. 124–25, 130, Sindall pp. 232, 234–5, T.I. Williams, *A Short History of 20th Century Technology,* pp. 287–8.

10 *World's Paper Trade Review,* 22 Feb., 6, 13 Dec. 1907. The fibrous raw materials used in papermaking at Mugiemoss are examined in Chapter 8, section 1.

11 DA, *Annual Report and Accounts,* 1910, 1913, 1914, Private Journal No. 2, 1895–1914, pp. 54, 57, 61–2. The Journal recorded a combined sum for Health and Unemployment Insurance of £52 in the financial year 1911–12 and more than £240 a year in the following two years.

12 See for example Pollard, pp. 42, 76, Taylor, pp. 28–9, 47–8.

13 Keith, p. 527, *Aberdeen Press and Journal* 23 Aug. 1951.

14 DA, *Annual Report and Accounts,* 1915–1918.

15 *Paper-maker,* Jan.–June 1917, pp. 18, 291, July–Dec. 1917, p. 541, Annual Number 1917–18 pp. 25, 26, 47.

16 *Paper-maker,* Annual Number 1917–18, pp. 26–30, Jan.–June 1919, pp. 259, 653. For some examples of paper-making firms which experienced a substantial increase in profits during the war see L. Weatherill, *One Hundred Years of Papermaking,* pp. 82–4, M. Tillmanns, *Bridge Hall Mills,* p. 91, N. Watson, *The Last Mill on the Esk,* pp. 113–14.

17 DA, *Annual Report and Accounts,* 1914–1918, Consumption of

Coals, Mugiemoss, monthly summaries 1895–1923 (with annual paper production).

18 Pollard, p. 64 and Table 14 below.

19 *Paper-maker*, Jan.–June 1920, pp. 32C, 32D, 755, Annual Number 1919–20, pp. 38–9, D.H. Aldcroft, *The Inter-War Economy: Britain 1919–1939*, pp. 31–6.

20 DA, *Annual Report and Accounts*, 1919, 1920, Private Journal No. 3, 1915–1932, pp. 1–19.

21 Aldcroft, p. 37, *Paper-maker*, Jan.–June 1921, pp. 23, 163, Annual Number 1922, p. 67.

22 *Paper-maker*, Jan.–June 1921, pp. 23, 644E, 644G, 802, Jan.–June 1922, pp. 3, 491, July–Dec. 1922, p. 58.

23 DA, *Annual Report and Accounts*, 1921.

24 DA, *Annual Report and Accounts*, 1922.

25 Aldcroft, p. 37, *Paper-maker*, Jan.–June 1926, p. 13.

26 *Report of the Board of Trade Safeguarding of Industries Committee on Packing and Wrapping Paper* Cmd. 2539 (1925).

27 DA, Daily Report Book, 1923–25. The earlier output figure (for 1899) is derived from monthly summaries in the Directors' Minute Book, 1898–1900.

28 *Report of Committee on Packing and Wrapping Paper*, paragraph 25.

29 *Paper-maker*, July–Dec. 1923, pp. 638–9, 648A, 648B.

30 DA, *Circular letters to Shareholders*, 29 Sept. and 3 Oct. 1925.

31 DA, Letters of Resignation by Alexander Marr, W.D. Davidson, and A.J. Davidson, 6 April, 9 Sept. 1926.

32 S.R.O., Files of dissolved companies, Morrison's Economic Stores Ltd., Return of Directors 11 July 1924, Keith, p. 401, *Post Office Aberdeen Directory*, 1911–12, 1924–25. I am also indebted to Mrs A Maxwell for conversation when I was a fellow member of Cults Tennis Club about her memories of the Stores in her youth.

33 *Paper-maker*, Jan.–June 1926, pp. 342–3, 607, Jan.–June 1931, p. 341, Pollard, pp. 193–5.

34 *Paper-maker*, Jan.–June 1932, pp. 302, 418, 447, July–Dec. 1932, pp. 223, 332, Jan.–June 1933, pp. 482–4, DA, *Annual Report and Accounts*, 1932, 1933.

35 Aldcroft, pp. 32–4, 38–9, *Paper-maker*, Annual Number 1929, p. 9, Jan.–June 1931, p. 124, Jan.–June 1932, p. 24.

36 DA, *Annual Report and Accounts*, 1927, *Paper-maker*, Jan.–June 1927, p. 189, July–Dec. 1927, p. 51.
37 *The History of BPB Industries*, p. 85, DA, Private Journal No. 3, 1915–1932, pp. 36–7, 39–40.
38 PRO, Board of Trade, Files of dissolved companies, the Colonial Paper Co. Ltd., *Annual Balance Sheets*, 1912–1928, DA, Private Journal No. 3 1915–1932, pp. 26–42, Agreement with W.H. Wassall of Sydney 8 June 1928 and associated correspondence.
39 C.J. Bundock, *The Story of the National Union of Printing, Bookbinding and Paper Workers*, pp. 429–30.
40 DA, *Circular letters to Shareholders*, 7 July 1931.
41 DA, Internal memorandum on proposal to float the sales organisation as a separate company, Davidson's Paper Sales Ltd., Directors' Minute Book, 1931–62. See also CRO Edinburgh, Davidson Files, Prospectus for issue of 4.5 per cent Redeemable Mortgage Debentures, Nov. 1950. The finances of the subsidiary Sales Company were not entirely separated from that of the parent company until April 1938.
42 DA, *Circular letter to Shareholders*, 6 Aug. 1931.
43 DA, *Special Resolution passed at Extra-ordinary General Meeting*, 30 July 1931 (for circulation to shareholders). See also note 34 above.
44 DA, Paper-making Department, Wages Book, No. 3. The gross wages bill in the main paper-making department averaged some £170 a week in January and February 1931, declined to some £72–106 a week for much of the period 30 May–7 Nov. 1931 and then increased to £152 a week from 14 Nov.–5 Dec. and £191 a week over the next three weeks.
45 DA, *Annual Report and Accounts*, 1932–34, Davidson's Paper Sales Ltd., Directors' Minute Book, 18 Oct. 1932.
46 CRO Edinburgh, Davidson Files, Report by W.F.M. Whitelaw, w.s., on petition of the firm to the Court of Session for sanction of the Scheme of Arrangement and confirmation of the reduction in Capital 22 Nov. 1935. See also the Scheme of Arrangement.
47 *Paper-maker*, Jan.–June 1934, pp. 116–17, 249–50, July–Dec. 1934, pp. 226–27, Jan.–June 1935, pp. 31–2, 365–6.
48 CRO Edinburgh, Davidson Files, Petition to the Court of Session, November 1935.

49 S.R.O., Files of dissolved companies, Morrison's Economic Stores Ltd., Return of Directors 11 July 1924.
50 CRO Edinburgh, Davidson Files, Scheme of Arrangement, p. 2, Report by W.F.M. Whitelaw, pp. 6, 7. *BC* A.T. Dawson, 2 July 1885, *MC* A.T. Dawson, 12 April 1910, *Directory of Directors*, 1928, 1930.
51 CRO Edinburgh, Davidson Files, Report by W.F.M. Whitelaw, Scheme of Arrangement, Resolutions passed at Extra-ordinary General Meeting 30 Sept. 1935, DA, *Annual Report and Accounts*, 1935.
52 *Directory of Directors*, 1924, 1930, 1933.
53 *Ibid*. 1933.
54 DA, *Annual Report and Accounts*, 1936.
55 DA, Colonel Davidson's Letter Book, letter to William Graham Jnr. & Co., Lisbon, 28 Nov. 1940, *Paper-maker*, Annual Number 1952, p. 92.
56 *Concrete Year Book*, 1939, p. 165, R.A.B. Smith, *Design and Construction of Concrete Roads*, p. 98.
57 Smith, pp. 98–101, *Concrete Year Book*, 1939, pp. 22–3, 32–3, 146, 149–51.
58 *The History of BPB Industries*, p. 87, *Concrete Year Book*, 1939, p. 888, DA, Sales Manager's Report, Nov. 1938.
59 *Paper-maker*, Jan.–June 1936, p. 241.
60 See for example Sindall, pp. 59–60, 81–2.
61 *Paper-maker*, Jan.–June 1932, pp. 24, 418, Jan.–June 1933, p. 4.
62 *Paper-maker*, Jan.–June 1932, p. 447, Jan.–June 1934, p. 418. The regulations specified a ream containing 480 sheets of Double Crown measuring 30 inches by 20 inches.
63 *Paper-maker*, Annual Number 1932, pp. 2, 8–16B, Jan.–June 1933, pp. 520–1, July–Dec. 1934, p. 188, Jan.–June 1936, pp. 176–7, July–Dec. 1936, p. 462, Annual Number p. 4.
64 *The History of BPB Industries*, pp. 86–7, *World's Paper Trade Review*, Jan.–June 1950, p. 881, *Directory of Paper-makers of the UK*, 1935. Producers of other types of board listed in the Directory were also much fewer in Scotland.
65 DA, *Annual Report and Accounts*, 1937, 1938, Daily Report Books, 1936–1938, *Paper-maker*, Jan.–June 1938, pp. 248, 391, Annual Number, p. 6, *World's Paper Trade Review*, July–Dec. 1938, pp. 2 2048, 2050, 2052. The output data is summarised in Chapter 4, Table 20.

66 CRO Edinburgh, Files of Northern Waste Paper Co. Ltd., *Memorandum and Articles of Association*, Return of Directors 18 Oct. 1937, Notification of change of registered office, March 1939, DA, Lists of waste paper contracts Sept. 1938–May 1939, Letter to Company Secretary 21 July 1939.
67 DA, Daily Record Books, 1935–37, 1937–39.
68 M. Bowley, *Innovations in Building Materials*, pp. 327, 330–2. The BPB operations are recorded in more detail in the official history previously cited.
69 *The History of BPB Industries*, p. 87, DA, Daily Report Book, 1937–39, Lists of Board Contracts outstanding Feb. 1938–May 1939.
70 See Tables 19 and 20 in Chapter 4 for the actual tonnage of paper and millboard produced.
71 DA, *Annual Report and Accounts*, 1936, 1937. See also Appendix C.
72 DA, *Annual Report and Accounts*, 1938.
73 DA, *Annual Report and Accounts*, 1939.

Chapter 4 The War Years, 1939–1945

1 S. Pollard, *The Development of the British Economy, 1914–1950*, pp. 297, 300.
2 *Ibid.* pp. 297–8, *Paper-Maker*, July–Dec. 1939, p. 256.
3 *Paper-Maker*, July–Dec. 1939, pp. 232, 234.
4 *Paper-Maker*, Jan.–June 1940, p. 138.
5 *Paper-Maker*, Jan.–June 1940, pp. 138, 140.
6 *Paper-Maker*, July–Dec. 1939, pp. 234–42, 304–5.
7 J. Hurstfield, *The Control of Raw Materials*, p. 179, *Paper-Maker*, Jan.–June 1940, p. 146.
8 *Paper-Maker*, Jan.–June 1940, p. 146.
9 A.J.P. Taylor, *English History 1914–1945*, pp. 560–2.
10 Hurstfield, p. 123, *Paper-Maker*, Jan.–June 1940, p. 138.
11 *Paper-Maker*, July–Dec. 1941, pp. 280–6, July–Dec. 1943, p. 34.
12 *Paper-Maker*, Jan.–June 1940, p. 220, July–Dec. 1941, p. 12, Annual Number 1942, p. 44.
13 *Paper-Maker*, July–Dec. 1942, pp. 62, 103, 188, July–Dec. 1943, p. 214.

14 *Paper-maker*, Jan.–June 1941, pp. 108, 168, Jan.–June 1942, pp. 59, 144A, 232, July–Dec. 1942, p. 62.
15 *Paper-maker*, Jan.–June 1941, pp. 3, 170.
16 W.J. Reader, *Bowater: a history*, p. 164, Hurstfield p. 69.
17 *Paper-maker*, Jan.–June 1942, pp. 5, 8, Jan.–June 1943, pp. 148–9.
18 Central Statistical Office, *Statistical Digest of the War*, p. 123, *Paper-maker*, Annual Number 1937, p. 38.
19 *Statistical Digest of the War*, p. 123.
20 Reader, pp. 157–62, *Statistical Digest of the War*, p. 123.
21 *Paper-maker*, Jan.–June 1942, p. 5, *Statistical Digest of the War* p. 123.
22 *Statistical Digest of the War*, p. 123.
23 DA, Colonel Davidson's Letter book, letters to various Branches, 28 May 1941.
24 *Paper-maker*, July–Dec. 1939, p. 240, C.R.O. Edinburgh, Files of Northern Waste Paper Co Ltd, Return of Directors 27 Oct. 1939, Return of Shareholders and Capital 28 Oct. 1939.
25 DA, Colonel Davidson's Letter book, letters to London Branch 29 Jan., Edinburgh Branch 2 April 1941.
26 *Paper-maker*, July–Dec. 1941, p. 108, DA, *Annual Report and Accounts*, 1942.
27 Keith, pp. 531–2, DA, Daily Report Book 1939–1941, entries for 17, 20 July 1940, Colonel Davidson's Letter Book, letters of 18 Sept., 12, 15, 23, 25 Oct., 2, 14 Nov., 21 April, 7, 9, 10, 13 May 1941.
28 DA, Colonel Davidson's Letter Book, letters to The Paper Control, Reading, 19, 24 Aug. 1940, to the Liverpool Branch 16 Dec. 1940, Daily Report Book 1939–1941, entries for May 1940.
29 DA, Daily Report Book 1939–1941.
30 DA, Colonel Davidson's Letter Book, letters of 31 Aug. 1940.
31 *Ibid.* letters of 26 Feb., 3 April, 1941, DA, Daily Report Book 1939–1941, entries for Nov. and Dec. 1940.
32 DA, Colonel Davidson's Letter book, letters of 14 Oct. 1940, 7, 11, 22, 23, 27 Jan., 4 April, 12 May 1941.
33 *Ibid.* letters of 15, 20, 22, 27, 28 Aug., 11 Sept., 4, 30 Oct., 14 Nov. 1940, DA, Daily Report Book 1939–1941.

34 DA, Colonel Davidson's Letter book, letters to London Branch and to Appleby and Wood, Manchester 25 Nov., 16 Dec. 1940, 27 Jan., 28 April 1941.
35 *Ibid.* letters of 18, 25 Nov., 13 Dec. 1940, 23 Jan., 4 Feb., 3, 12 March 1941.
36 DA, *Annual Report and Accounts*, 1940–1945.
37 DA, *Annual Report and Accounts*, 1936–1945.
38 C.R.O. Edinburgh, Davidson Files, Returns of Allotments of Shares by conversion of 5 per cent convertible mortgage debentures at various dates 1 Feb. 1943–31 May 1945, DA, *Annual Report and Accounts*, 1941, 1943–1945.
39 DA, *Annual Report and Accounts*, 1945.
40 DA, *Annual Report and Accounts*, 1942, 1945, *Paper-maker*, July–Dec. 1944, pp. 214, 262, Annual Number p. 22.

Chapter 5 Post War Expansion and New Horizons, 1945–1953

1 Will of A.T. Dawson, Parish of St. Brelade, Jersey, 23 May 1962, C.R.O. Edinburgh, Davidson Files, Return of Directors 18 Sept. 1946, *Annual Report and Accounts*, 1946.
2 C.R.O. Edinburgh, Davidson Files, Returns of Directors 18 Sept. 1946, DA, Davidson's Paper Sales Ltd., Directors Minute Book, Report of Annual General Meeting 28 Dec. 1938, *Stock Exchange Official Year Book*, 1947, *The History of BPB Industries*, p. 88.
3 C.R.O. Edinburgh, Davidson Files, Returns of Directors, 9 Aug. 1948, DA, *Annual Report and Accounts*, 1943–1945.
4 C.R.O. Edinburgh, Davidson Files, *Annual Report and Accounts*, 1946, DA, *Annual Report and Accounts*, 1943.
5 See for example J.C.R. Dow, *The Management of the British Economy 1945–60*, pp. 149–62, Pollard pp. 356–62.
6 *Paper-maker*, Jan.–June 1948, p. 185.
7 *Annual Abstract of Statistics*, 1955 Table 182, 1959 Table 181, *Paper-maker*, Jan.–June 1947, p. 72, July–Dec. 1947, p. 151, Jan.–June 1948, p. 185, July–Dec. 1948, p. 8.
8 R.L. Hills, *Papermaking in Britain 1488–1988*, p. 188, *Annual Abstract of Statistics*, 1955 Table 182, 1959 Table 181.
9 DA, Minutes of Board Meetings 1948–1952, Report approved

30 May 1950, C.R.O. Edinburgh, Davidson Files, *Annual Report and Accounts*, 1951.

10 DA, Minutes of Board Meetings 1948–1952, acceptance of Walmsley's tender 3 Feb. 1949, Report approved 30 May 1950.

11 *Ibid.* Report approved 30 May 1950.

12 *HC Debates*, 5th series, vol. 481, cols. 668, 669, 682–5, 696–7, 24 Nov. 1950, Dow, p. 55.

13 *Paper-maker*, Jan.–June 1951, pp. 38, 236, 251, July–Dec. 1951, pp. 6, 384, Jan.–June 1952, p. 281.

14 *HC Debates*, 5th series, vol. 481, cols. 446–54, 2 Nov. 1950.

15 C.R.O. Edinburgh, Davidson Files, *Annual Report and Accounts*, 1951.

16 C.R.O. Edinburgh, Davidson Files, *Annual Report and Accounts*, 1946, *Prospectus for issue of four and a half per cent Redeemable Mortgage Debentures*, Nov. 1950. DA, Minutes of Board Meetings 1948–1952, 29 June 1948, 28 June, 9 Sept. 1949, 28 March 1952, *the History of BPB Industries*, p. 94.

17 *Newcastle Journal*, 7 Jan. 1937, *Industrial Estates: a Story of Achievement*, pp. 166–169, C.R.O. Edinburgh, Davidson Files, *Prospectus for issue of four and a half per cent Redeemable Mortgage Debentures*, Nov. 1950, Agreement with the Directors of Fibreboard Boxes Ltd., 23 Jan. 1949, DA, Minutes of Board Meetings 1948–1952, 25 May, 29 June 1948.

18 C.R.O. Edinburgh, Davidson Files, Agreement registered 6 June 1946.

19 Sir Eric Bowater's career is examined at length in W.J. Reader's *Bowater: a history*.

20 DA, Minutes of Board Meetings 1948–1952, 27 June, 26 Oct. 1948.

21 *Ibid.* 22 Dec. 1948, 22 Feb. 1949.

22 *Ibid.* 13, 29 March, 17 Oct. 1949.

23 *Ibid.* 24 Jan. 1950, *World's Paper Trade Review*, 11 Nov. 1938, *Stubbs' Directory: Manufacturers, Merchant Shippers and Professional*, 1950–1951.

24 DA, Minutes of Board Meetings 1948–1952, 28 Feb., 25 April, 30 May 1950.

25 *Ibid.* 28 Sept., 26 Oct. 1948, 3, 22 Feb., 31 May 1949.

26 *Ibid.* 31 May, 28 June 1949.

27 *Ibid.* 30 Aug., 25 Oct., 29 Nov. 1949.

28 *Ibid.* 31 Jan., 28 March, 30 May 1950.
29 *Ibid.* 30 May, 27 June, 31 Oct., 26 Dec. 1950.
30 C.R.O. Edinburgh, Davidson Files, *Prospectus for issue of four and a half per cent Redeemable Mortgage Debentures*, Nov. 1950.
31 For a convenient summary see W.A. Thomas, *The Finance of British Industry*, 1918–1976, pp. 229–31.
32 C.R.O. Edinburgh, Davidson Files, Resolution passed at Annual General Meeting 13 Nov. 1945, *Annual Report and Accounts*, 1946.
33 DA, Minutes of Board Meetings 1948–1952, 25 May 1948.
34 *Ibid.* 17 Aug. 1948.
35 *Ibid.* 17, 27 Aug., 10 Sept., 29 Oct. 1948, C.R.O. Edinburgh, Davidson Files, Resolutions passed at Extra-ordinary General Meetings 15 Feb. 1946, 10 Sept. 1948.
36 DA, Minutes of Board Meetings 1948–1952, 29 Dec. 1949.
37 *Ibid.* 29 Nov., 29 Dec. 1949, 24 Jan. 1950.
38 *Ibid.* 28 Feb. 1950.
39 *Ibid.* 30 May, 20, 27 June, 21 July, 3, 17, 29 Aug., 5, 26 Sept. 1950.
40 *Ibid.* 17 Oct., 8, 28 Nov. 1950.
41 C.R.O. Edinburgh, Davidson Files, *Annual Report and Accounts*, 1946–1950, *Prospectus for issue of four and a half per cent Redeemable Mortgage Debentures*, Nov. 1950.
42 C.R.O. Edinburgh, Davidson Files, *Annual Report and Accounts*, 1951.
43 *Aberdeen Press and Journal*, 23 Aug. 1951, 4 Sept. 1986, DA, *Annual Report and Accounts*, 1911. For a brief survey of the structure of British manufacturing industry shortly before the First World War see L. Hannah, *The Rise of the Corporate Economy*, pp. 25–7.
44 C.R.O. Edinburgh, Davidson Files, *Annual Report and Accounts*, 1951, *Directory of Directors*, 1950, *Stock Exchange Official Year Book*, 1951, Dod's *Parliamentary Companion*, 1952.
45 DA, Minutes of Board Meetings, 1948–1952, 24, 31 Jan., 7, 28 Feb. 1950.
46 *Ibid.* 15 March, 30 Aug., 9 Sept., 17 Oct., 29 Nov., 6 Dec. 1949, *Paper-maker*, July–Dec. 1950, pp. 80, 320.
47 *Directory of Directors*, 1950, DA, Minutes of Board Meetings 1948–1952, 28 Dec. 1951, C.R.O. Edinburgh, Davidson Files, Annual Return of Share Capital and Directors, 13 Nov. 1953.
48 Bowley, p. 336.

49 DA, Minutes of Board Meetings, 1948–1952, undated minute of summer 1951 under heading 'Gyproc Ltd'. *The History of BPB Industries*, pp. 62, 66–7, 88–9.
50 DA, Minutes of Board Meetings, 1948–1952, 19 Oct., 28 Dec. 1951, undated minute of Autumn 1951, 25 Jan., 29 Feb. 1952– all items headed 'Rhodesian Project'.
51 *The History of BPB Industries*, p. 89.
52 *Aberdeen Press and Journal*, 2 April 1953.
53 *Paper-maker*, Jan.–June 1952, p. 465, July–Dec. 1952, pp. 5, 285.
54 *Ibid.* July–Dec. 1952, pp. 24, 471, Jan.–June 1953, p. 3, *Annual Abstract of Statistics*, 1955, Table 182.
55 See Tables 19 and 20 in Chapter 4 as well as Tables 21 and 22.
56 DA, Minutes of Board Meetings 1948–1952, 30 May 1950.
57 C.R.O. Edinburgh, Davidson Files, *Annual Report and Accounts*, 1952. Most of the data is summarised in Tables 23 and 24 above.
58 C.R.O. Edinburgh, Davidson Files, *Report and Accounts*, 2 Aug. 1952–31 March 1953.
59 DA, Minutes of Board Meetings 1948–1952, 28 Aug. 1951, 29 Feb. 1952.
60 C.R.O. Edinburgh, Files of Abertay Paper Sacks Ltd., *Annual Report and Accounts*, year ending 30 Sept. 1952.
61 C.R.O. Edinburgh, Davidson Files, *Report and Accounts*, 2 Aug. 1952–31 March 1953.
62 *Ibid.* Resolution passed at Extra-ordinary General Meeting 26 Sept. 1952.
63 *Aberdeen Press and Journal*, 2 April 1953, C.R.O. Edinburgh, Davidson Files, *Report and Accounts*, 2 Aug. 1952–31 March 1953.
64 *Aberdeen Press and Journal*, 2 April 1953.
65 *Times*, 21 April, 14 May 1953, *Economist*, 1 Aug. 1953, C.R.O. Edinburgh, Davidson Files, *Report and Accounts* 2 Aug. 1952–31 March 1953. The Annual General Meeting was held on 30 Oct. 1953 and the Directors' Report and Chairman's Statement both referred to the British Plaster Board Ltd take-over.

Chapter 6 The Shareholders

1 DA, *Memorandum and Articles of Association*, 1875, C.R.O. Edinburgh, Davidson Files, Annual Return of Shareholders and Capital, 18 Oct. 1877.
2 C.R.O. Edinburgh, Annual Return of Shareholders and Capital, 10 Oct. 1883, 10 Oct. 1891.
3 The Wages Censuses of 1886 and 1906 reveal that the most prosperous 10 per cent of adult male manual workers earned more then 34s 7d per week in 1886 rising to more than 46s 0d in 1906. The least prosperous 10 per cent earned less than 16s 7d per week in 1886 and less than 19s 6d per week 20 years later. A.L. Bowley, *Wages and Income in the U.K. since 1860*, p. 42.
4 *A Century of Papermaking, 1820–1920, Robert Craig & Sons Ltd.*, pp. 14, 59, 65–6, 74.
5 See for example E.H. Hunt, *British Labour History, 1815–1914*, pp. 17–25. S. Buckley 'The Family and the Role of Women', *The Edwardian Age Conflict and Stability 1900–1914* ed. A. O'Day, pp. 136–7.
6 Cottrell, pp. 91–6.

Chapter 7 Owners and Directors

1 The names of the Directors are listed together with their year of birth and death and their period in office in Appendix A.
2 C.W. Davidson, W.D. Davidson and A.J. Davidson each had a son who might have become a director in different circumstances. See respectively *World's Paper Trade Review*, 21 May 1943, DC, Aberdeen, 20 Aug. 1940, (and BC of his son N.D. Davidson, Aberdeen, 10 Aug. 1910), DC, Wimbledon, 17 April 1966.
3 *Aberdeen Press and Journal*, 4 Sept. 1986.
4 'Mr Alexander Davidson, Mr John Davidson and Mr David Davidson', *Men of the Period, Scotland, Paper-maker*, 2 Aug. 1920, *World's Paper Trade Review*, 21 May 1943, C.R.O. Edinburgh, Davidson Files, Annual Return of Shareholders and Capital, 10 Oct. 1891, *Aberdeen Press and Journal*, 23 Aug. 1951. The two sons of Alexander Davidson who became directors followed a different educational path, being born and brought

up in London where their father was in charge of the London Office.

5 *Paper-maker*, July–Dec. 1951, p. 156.

6 BC, W.D. Davidson and A.J. Davidson, Kensington, 18 Oct. 1871, 13 Jan. 1876, C.W. Davidson, Newhills, Aberdeenshire, 15 June 1869, *Aberdeen Press and Journal*, 4 Sept. 1986. See also Notes 2 and 4 above.

7 S.R.O., Sheriff Court Books for Aberdeenshire, Inventory of John Davidson, 6 Dec. 1897, DA, Register of Directors, 1901–1936, *Directory of Directors*, 1909, 1910, C.R.O. Edinburgh, Davidson Files, Return of Directors, 19 Dec. 1952.

8 S.R.O., Sheriff Court Books for Aberdeenshire, Inventories of John Davidson, 6 Dec. 1897, David Davidson, 10 May 1915.

9 He left £16,288 and after his death his executors held shares in the firm with a nominal value of less than £5,000. Probate Calendar, Principal Register of the Family Division, Somerset House, probate granted 24 Aug. 1920, DA, Share Ledger, 1923–1945.

10 *Aberdeen Daily Journal*, 15 Feb. 1915, 12 July 1920, *Aberdeen Press and Journal*, 23 Aug. 1951, 4 Sept. 1986.

11 *Paper-maker*, July–Dec. 1920, p. 172A, July–Dec. 1951, p. 156.

12 *Aberdeen Press and Journal*, 23 Aug. 1951, 4 Sept. 1986, *Paper-maker*, July–Dec. 1950, p. 150, *London Gazette*, April–June 1945, p. 3376.

13 C.R.O. Edinburgh, Davidson Files, Return of Directors, 9 Aug. 1948, Return of Directors, 8 Jan. 1952. See also notes 14, 17–19, 23 below.

14 *World's Paper Trade Review*, 4 Aug. 1911, C.R.O. Edinburgh, Davidson Files, Annual Return of Shareholders and Capital 18 Oct. 1877, 10 Oct. 1883, 10 Oct. 1891, DA, Return of Salaries at London Office (for submission to the Inland Revenue) 4 July 1893.

15 DA, *Annual Report and Accounts*, 1943.

16 See Ch. 5, notes 3 and 45.

17 DC, Rubislaw, Aberdeen, 8 Feb. 1930, *Aberdeen Press and Journal*, 11 Feb. 1930. See also Ch. 3, note 1.

18 *Dunfermline Press and West of Fife Advertiser*, 8 April 1982, Times, 10 April 1982, *Directory of Directors*, 1933, BC, Govan, 6 Feb. 1893.

19 *Cambuslang Advertiser*, 5 Jan. 1957, BC Ochiltree, Ayshire, 12 April 1877.

20 *Aberdeen Press and Journal*, 2 Nov. 1970, *Directory of Directors*, 1933, 1945, BC, St. Machar, Aberdeen 25 Jan. 1909.

21 *Who's Who*, 1981, *Times*, 11 Sept. 1981, *Dod's Parliamentary Companion*, 1951, *Stock Exchange Official Year book*, 1951, C. Gulvin, *The Scottish Hosiery and Knitwear Industry, 1680–1980*, p. 97. His Constituency in 1945 was the Central Division, County of Aberdeen and Kincardine, and from 1950 to his retirement in 1959 West Aberdeenshire.

22 N.J. Morgan, 'William Tawse', in *Dictionary of Scottish Business Biography 1860–1960*, ed. A Slaven and S. Checkland, vol. 2, *Aberdeen Press and Journal*, 5 July 1972, 20 Nov. 1980.

23 BC, Buckie 2 July 1885, MC, St. Machar, Aberdeen 12 April 1910, S.R.O., Files of dissolved companies, Morrison's Economic Stores Ltd. Change of name and company headquarters 20 June 1924, Return of Shareholders and Directors 11 July 1924, Agreements between James Mearns and the company 20 May 1926 (with retrospective effect). The company legally began its existence as Brooks and Hamilton, a Glasgow based private company registered in 1908. Morrison's Economic Stores had existed in Aberdeen for some years before A.T. Dawson joined the firm after his marriage.

24 *Aberdeen Press and Journal*, 18 Aug. 1962, BC, Aberdeen, 6 Aug. 1891.

25 *Aberdeen Press and Journal*, 1 March 1969, *Directory of Directors*, 1950. See also ch. 5 note 47.

26 *Cambuslang Advertiser*, 5 Jan. 1957.

27 *Aberdeen Press and Journal*, 18 Aug. 1962, 2 Nov. 1970, *Dunfermline Press and West of Fife Advertiser*, 8 April 1982.

28 *Aberdeen Press and Journal*, 1 March 1969, 20 Nov. 1980, *Cambuslang Advertiser*, 5 Jan. 1957, *Who's Who*, 1981. See also note 27 above.

29 *Aberdeen Press and Journal*, 18 Aug. 1962, *Dunfermline Press and West of Fife Advertiser*, 8 April 1982.

30 A. Slaven's, 'Conclusions: Some Characteristics of Scottish Entrepreneurs', *Dictionary of Scottish Business Biography*, vol. 2, ed. A. Slaven and S. Checkland, p. 430. The university graduates were Duffus, Ledingham, McCrone, Mellis and Tawse.

Chapter 8 The Labour Force

1 Coleman, pp. 26, 108, 337–8.
2 Coleman, p. 343, Hills, pp. 138–140, *Annual Statement of Trade of the United Kingdom*, P.P., 1863–1890. Until 1868 the customs enumerators lumped together esparto with other vegetable fibres.
3 Hills, pp. 146–152, *Annual Statement of Trade of the United Kingdom*, 1869–1890. In the years 1871–1886 wood pulp imports were not distinguished separately from those of rag pulp. However the latter were probably very small, never exceeding 190 tons a year in the three years 1868–1870, for example, when they were recorded separately. In 1870 imports in the category 'Old Ropes or Junk, old fishing nets, and other materials for making paper', which included woodpulp, amounted to 5268 tons, of which 2590 tons came from Norway and Sweden. In 1869 the total figure in this category was only 1663 tons.
4 Spicer, pp. 29–32, 51–2, 183–5, 202–3, 217–21, Tillmanns, pp. 63–4, 75, 78.
5 *Fourth Report of the Commissioners appointed in 1868 to inquire into the best means of preventing the pollution of rivers. Pollution of Rivers in Scotland*, PP 1872, XXXIV, Evidence, Answers to Queries, Part ii, 225.
6 DA, Directors' Minute Book 1898–1900, 31 May, 4 Sept. 1899, Letter from H.M. Paterson, HM Inspector of Factories, 11 Oct. 1901.
7 AUA, MS 2769, I/30/1, Minute Book 1871–1875, pp. 10, 14, Spicer, p. 30.
8 *World's Paper Trade Review*, 18 April 1913, AUA, MS 2769, I/30/1, Minute Book 1871–1875, pp. 6, 9.
9 AUA, MS 2769, I/30/1, Minute Book 1871–1875, pp. 10, 13–14, DA, Inventory of Title Deeds, 1877, lease of Size Yard, London, 3 June 1872, *Paper Mills Directory*, 1890, 1895.
10 DA, *Annual Report and Accounts*, 1916. See also Chapter 3, Section 3, text and note 29.
11 See Chapter 3, Section 5.
12 Dawe, pp. 8–9, Sindall, pp. 13–15.
13 Dawe, pp. 9–11, Sindall, pp. 17–20, 47–52, 79.
14 Dawe, pp. 20–1, Spicer, pp. 81–4, Sindall, pp. 20, 40–2, Cross and Bevan, pp. 183–5, 196–7.

15 Dawe, pp. 13, 36, Sindall, pp. 70, 79.
16 Dawe, pp. 21–23, Sindall, pp. 26–8.
17 Dawe, p. 21, Hills pp. 157–8, Spicer, pp. 64–6. See also chapter 2, Note 40.
18 Dawe, pp. 23–5, Sindall, p. 31.
19 Dawe, pp. 25, 20–32, Sindall, pp. 25, 34.
20 Dawe, pp. 32–3, Sindall, pp. 34–5.
21 Coleman, p. 290, *New Statistical Account of Scotland*, XII, 72, 239.
22 See Chapter 1, Note 32.
23 *Aberdeen Journal*, 7 Jan. 1896, 'Mr Alexander Davidson, Mr John Davidson, and Mr David Davidson,' *Men of the Period, Scotland.*
24 DA, Quotation by Royal Scottish Insurance Co. Ltd. for insurance of workers at Mugiemoss in connection with the Worker's Compensation Act 1906, June 1908.
25 DA, Paper-making Department, Wages Book No 3, week ending 15 Feb. 1930, *Aberdeen Press and Journal*, 8 Jan 1937.
26 *World's Paper Trade Review*, Jan–June 1950, p. 881.
27 DA, Quotation for insurance of workers at Mugiemoss, June 1908.
28 J.N. Bartlett, *Carpeting the Millions : The growth of Britain's Carpet Industry*, p. 115, E.H. Phelps Brown and Margaret H. Browne, *A Century of Pay*, p. 126.
29 DA, Paper-making Department, Wages Book No 3, week ending 15 Feb. 1930.
30 *Royal Commission on Labour : Textile, Clothing, Chemical, Building and Miscellaneous Trades*, PP 1893–94, XXXIV, 763, 769, Dawe, p. 29, Sindall, p. 14. See also Watson pp. 55–6, Weatherill, pp. 34–6.
31 A.H. Shorter, *Paper-making in the British Isles*, p. 216. I calculated the percentages. The Donside Paper Co. was not founded until 1893, replacing the ailing Gordon's Mills Paper Co. Ltd. on the same site.
32 Census of Scotland, 1891, Population II, PP 1893–94, CVIII
33 *Royal Commission on Labour*, PP 1893–94, XXXIV, 763.
34 A. Fowler, 'Quality depends on the Beaterman', *Paper Worker*, I, pp. 20–1. See also Watson, p. 58.
35 A. Fowler, 'Why Paper is not Always Perfect', *Paper Worker*, I, pp. 27–8, Sindall, p. 31.

36 Spicer, p. 66, *Children's Employment Commission (1862): Fourth Report of the Commissioners*, PP 1865, XX, 172–5, Tillmanns, pp. 84–5.
37 *Report of an Enquiry by the Board of Trade into the Earnings and Hours of workpeople of the U.K.*, VIII, *Paper, Printing etc Trades in 1906*, PP 1912–13, CVIII, pp. 325–48.
38 A.L. Bowley, *Wages and Income in the U.K. since 1860*, p. 42.
39 DA, Quotation for insurance of workers at Mugiemoss, June 1908.
40 New Register House, Edinburgh: General Register Office, Census Enumerators Books for Newhills Parish, Aberdeenshire, 1851–1891.
41 S.R.O., Register of Sasines, Aberdeenshire, 20 March 1871, 2 Feb. 1871, DA, Annual Report and Accounts, 1886–1935, Inventory of Title Deeds, 1914, Greenburn Houses and Mugiemoss Cottages.
42 DA, Letter Book of John Davidson, 14 Oct. 1880, 17 April, 1 May 1884. See also S.R.O., Valuation Rolls for Aberdeenshire. However the Davidson tenants in the Roll for 1892–93, for example, included labourers as well as mill managers, engineers and millwrights.
43 DA, Minutes of Board Meetings, 1948–1952, 29 Dec. 1949, 31 Jan., 25 April, 30 May, 27 June 1950.
44 Keith, pp. 493–4, M.J. Mitchell and I.A. Souter, *The Aberdeen District Tramways*, p.15.
45 M.J. Mitchell and I.A. Souter, *the Aberdeen Suburban Tramways*, pp. 8–18, 58.
46 H.A. Vallance, *The Great North of Scotland Railway*, pp. 25, 100, Appendix 4, Bradshaw's *Railway Guide*, 1887.
47 *Paper-Maker*, 16 Sept. 1901.
48 Hunt, p. 78.
49 B.L. Hutchins and A. Harrison, *A History of Factory Legislation*, pp. 38–42, 168–9, 233–4, Hunt, p. 12.
50 Hutchins and Harrison, pp. 105–12.
51 Hunt, p. 78.
52 *Report of an Enquiry into Earnings and Hours*, VIII, *Paper, Printing etc. Trades*, PP 1912–13, CVIII, 348.
53 *Royal Commission on Labour*, PP 1893–94, XXXIV, 522, 763.
54 *Hunt*, pp. 79–80, *Royal Commission on Labour*, PP 1893–94, XXXIV, 763, Watson, p. 63, Bundock, pp. 402–3.

55 Bundock, pp. 413–4, 527.
56 *Report of an Enquiry into Earnings and Hours*, VIII, *Paper, Printing etc. Trades*, PP 1912–13, CVIII, 348, Bundock, p. 453.
57 *Returns of Wages published between 1830 and 1886*, PP 1887, LXXXIX, 576, *Royal Commission on Labour*, PP 1893–94, XXXIV, 763.
58 DA, Paper-Making Department, Wages Book No. 3. See also Appendix C.
59 *Children's Employment Commission (1862), Fourth Report of the Commissioners*, PP 1865, XX, 142–3, 147–9, Dawe, pp. 23–4, 29, Weatherill, pp. 34–6, *Report of the Chief Inspector of Factories*, 1901, PP 1902, XII, 244, *Spicer*, p. 66, *Royal Commission on Labour*, PP 1893–94, XXXIV, 769.
60 *Royal Commission on Labour*, PP 1893–94, XXXIV, 523.
61 *Report of the Chief Inspector of Factories*, 1894, PP 1895, XIX, 11, 297, 1895, PP 1896, XIX, 441, 1898 Part I, PP 1899, XII, 7, 1899, PP 1900, XI, Prosecutions in Aberdeen District, DA, Letter from H.M. Paterson, HM Inspector of Factories and reply by company secretary, 11, 14 Oct. 1901.
62 Hunt, pp. 43–4.
63 Spicer, pp. 167–8.

Bibliography

Manuscript Collections

C. Davidson & Sons Ltd.
(Note: The archives of the firm were surveyed in the mid 1970's by the National Register of Archives, North Eastern Survey, Survey No. 1460. However the bulk of the material listed below was not recorded in the Survey. On the other hand a number of the manuscripts which were listed appear to have vanished and have not come to light despite repeated searches and enquiries within the firm with the full co-operation of the management. The most serious loss was the directors' minute books covering the years 1875–1934 of which only one volume covering the years 1898–1900, has survived. All the manuscripts listed below have now been deposited by the firm on permanent loan in Aberdeen University Archives.)

Annual Report and Accounts, 1875–1945 (in one volume)
Annual Return of shareholders, 1924–1945 (in one volume)
Circular Letters to Shareholders, 1891–1931
Colonel Davidson's Letter Book, 1940–1941
Daily Report Books 1923–1925, 1935–1953
Directors' Minute Book, Davidson's Paper Sales Ltd., 1931–1962
Inventory of Title Deeds, 1877
Inventory of Title Deeds, 1914
Letter Book of John Davidson, 1871–1900
Memorandum and Articles of Association, 1875 amended by special resolutions passed at various dates 1883–1918
Memorandum and Articles of Association, Boulinikon Felt Co. Ltd., 1883
Minutes of Board Meetings, 1948–1952

Miscellaneous letters, invoices and agreements, internal lists, memoranda, and reports (specified individually in the End Notes)
Pamphlet to celebrate the firm's centenary
Papermaking Department, Wages Book No. 3, 1930–1932
Private Journal No. 2, 1895–1914
Private Journal No. 3, 1915–1932
Private Ledger, 1852–1865
Prospectus for issue of four and a half per cent debentures, 1896
Register of Directors, 1901–1936
Register of Mortgages, 1876–1925
Share Ledger, 1923–1935

Other Collections
Aberdeen University Department of Special Collections and Archives:
Davidson and Garden MSS (2769)
Press Cuttings on Strike at Mugiemoss in 1936 (2649)
Companies Registration Office, Edinburgh
Files of returns to the Registrar of Companies:
Abertay Paper Sacks Ltd.
C. Davidson & Sons Ltd.
Northern Waste Paper Co. Ltd.
Judicial Greffe, St. Helier, Jersey:
Wills
New Register House, Edinburgh: General Register Office:
Birth, Death and Marriage Certificates
Census Enumerators' Books, Newhills, Aberdeenshire 1851–1891
Office of Population Censuses and Surveys, General Register Office, St Catherine's House, London:
Birth, Death and Marriage Certificates
Public Record Office, London:
Board of Trade, Files of Dissolved Companies (B.T.31): Colonial Paper Co. Ltd.
Scottish Record Office, Edinburgh
Register of Sasines, Aberdeenshire
Sheriff Court books, Aberdeenshire
Files of Dissolved Companies (B.T.2)
Valuation Rolls, Aberdeenshire

Somerset House, London, Principal Registry of the Family Division:
Probate Calendar, Wills
Superintendent Registrar, St Helier, Jersey
Death Certificates

Official Publications
Annual Statement of the Trade and Navigation of the United Kingdom, 1863–1890
Annual Abstract of Statistics, 1955, 1959
British Patents for Inventions
Abridgements of Specifications, 1859–1898
Specifications, 1859–1898
Census of Scotland, 1891, *Population II*, PP 1893–94, CVIII
Central Statistical Office, Statistical Digest of the War, H.M.S.O. 1951
Children's Employment Commission (1862): *Fourth Report of the Commissioners*, PP 1865, XX
Fourth Report of the Commissioners appointed in 1868 to inquire into the best means of preventing the pollution of rivers. Pollution of Rivers in Scotland, PP 1872, XXXIV
House of Commons Debates, Fifth Series, vol. 481, November 1950
Hurstfield, J. (1953) *The Control of Raw Materials*, Official History of the Second World War, U.K. Civil Series, H.M.S.O.
London Gazette, 1945
Report of the Board of Trade, Safeguarding of Industries, Committee on Packing and Wrapping Paper, Cmd. 2539, 1925
Report of the Chief Inspector of Factories, PP 1895–1902
Report of an Enquiry by the Board of Trade into the Earnings and Hours of the Workpeople of the U.K. VIII, *Paper, Printing etc Trades in 1906*, PP 1912–13, CVIII
Returns of Wages published between 1830 and 1886, PP 1887, LXXXIX
Royal Commission on Labour Textile, Clothing, Chemical, Building and Miscellaneous Trades, PP 1893–94, XXXIV

Periodical Publications

(Years are indicated only where usage was confined to one or two years)
Aberdeen Daily Journal
Aberdeen Journal
Aberdeen Press and Journal
Bradshaw's *Railway Guide*, 1887
Cambuslang Advertiser, 1957
Concrete Year Book: Handbook, Directory and Catalogue of Concrete, 1939
Directory of Directors
Directory of Papermakers of the U.K.
Dod's *Parliamentary Companion*, 1951
Dunfermline Press and West of Fife Advertiser, 1982
Economist, 1953
The Grocer, 1863, 1877
Newcastle Journal, 1937
Papermaker and British Paper Trade Journal
Paper Mills Directory
Paper Trade Review
(continued as *World's Paper Trade Review*)
Post Office Aberdeen Directory
Stock Exchange Official Year Book
Stubb's Directory: Manufacturers, Merchant Shippers and Professional, 1950–51
Times
Who's Who, 1981

Books and Articles

(The place of publication has been omitted for works published in London)

Aldcroft, D.H. (1970) *The Inter-War Economy: Britain 1919–1939*.
Bartlett, J. N. (1978) *Carpeting the Millions: The Growth of Britain's Carpet Industry*, Edinburgh.
—— (1980) 'Alexander Pirie & Sons of Aberdeen and the expansion of the British Paper Industry, c. 1860–1914, in *Business History*, XXII.

—— (1982) 'Investment for survival: Culter Mills Paper company Ltd., 1865–1914' in *Northern Scotland*, V, Aberdeen.
Bowley, A.L., (1937) *Wages and Income in the U.K. since 1860*, Cambridge.
Bowley, M. (1960) *Innovations in Building Materials.*
Brown, E.H. Phelps and Browne, Margaret H. (1968) *A Century of Pay.*
Buckley, S. (1979) 'The Family and the Role of Women', in A. O'Day (ed.) *The Edwardian Age: Conflict and Stability, 1900–1914.*
Bundock, C.J. (1959) *The Story of the National Union of Printing, Bookbinding and Paper Workers*, Oxford.
A Century of Papermaking, 1820–1920, Robert Craig & Sons Ltd. Privately printed, Edinburgh, 1920
Coleman, D.C. (1958) *The British Paper Industry*, Oxford.
Cottrell, P.L. (1980) *Industrial Finance, 1830–1914.*
Cross, C.F. and Bevan, E.J. (1900) *A Century of Papermaking*, 2nd edition.
Darwin, B. (1945) *Robinsons of Bristol, 1844–1944*, privately printed, Bristol.
Dawe, E.A. (1914) *Paper and its uses.*
Dow, J.C.R. (1970) *The Management of the British Economy, 1945–1960*, Student's edition, Cambridge.
Edelstein, M. (1976) 'Realised rates of return on U.K. Home and Overseas Portfolio Investment in the Age of High Imperialism' in *Explorations in Economic History*, XIII.
Evely, R. and Little, I.M.D. (1960) *Concentration in British Industry*, Cambridge.
Fowler, A. (1940) 'Quality depends on the Beaterman' and 'Why Paper is not Always Perfect', *Paper Worker*, I. 1940–1941.
Gulvin, C. (1984) *The Scottish Hosiery and Knitwear Industry, 1680–1980*, Edinburgh.
Hannah, L. (1976) *The Rise of the Corporate Economy.*
Hills, R.L. (1988) *Papermaking in Britain 1488–1988.*
The History of BPB Industries, privately printed, 1973.
(Based on research by David Jenkins and on information supplied by many employees past and present, and friends of the company).
Hunt, E.H. (1981) *British Labour History, 1815–1914.*
Hutchins, B.L. and Harrison, A. (1966) *A History of Factory Legislation*, 3rd edition.

Industrial Estates: A story of Achievement, Published for North Eastern Trading Estates Ltd., 1952.

Jeffreys, J.B. (1977) *Business Organisation in Great Britain, 1856–1914*, New York.

Keith, A. (1972) *A Thousand Years of Aberdeen*, Aberdeen.

Men of the Period, Scotland, c. 1896

Mitchell, M.J. and Souter, I.A.(1983) *The Aberdeen District Tramways*, Dundee.

—— (1980) *The Aberdeen Suburban Tramways*, Dundee.

Morgan, N.J., (1990) 'William Tawse' in A. Slaven and S. Checkland (eds), *Dictionary of Scottish Business Biography*, 1860–1960, vol 2, Aberdeen.

Morgan, P. (1886) *Annals of Woodside and Newhills*, Aberdeen.

New Statistical Account of Scotland, by the ministers of the respective parishes under the superintendence of a committee of the society XII, Edinburgh, 1845.

Payne, P.L. (1980) *The Early Scottish Limited Companies, 1856–1895*, Edinburgh.

Pollard, S. (1962) *The Development of the British Economy, 1914–1950*.

Reader, W.J. (1981) *Bowater: A History*, Cambridge.

Shorter, A.H. (1971) *Paper-making in the British Isles*, Newton Abbot.

Sindall, R.W. (1906) *Paper Technology: An Elementary Manual*.

Slaven, A., (1990) 'Conclusions: Some Characteristics of Scottish entrepreneurs', in A. Slaven and S. Checkland (eds) *Dictionary of Scottish Business Biography, 1860–1960*, vol. 2., Aberdeen.

Smith, R.A.B. (1934) *Design and Construction of Concrete Roads*.

Spicer, A.D. (1907) *The Paper Trade: A descriptive and historical survey*.

Strachan, J., (1949) 'The Invention of wood pulp processes in Britain during the 19th Century', *Paper-Maker*, Annual Number.

Taylor, A.J.P. (1965) *English History, 1914–1945*, Oxford.

Thomas, W.A. (1978) *The Finance of British Industry, 1918–1976*.

Thomson, A.G. (1974) *The Paper Industry in Scotland*, Edinburgh.

Tillmanns, M. (1978) *Bridge Hall Mills. Three Centuries of Paper and Cellulose Film Manufacture*, Tisbury, Wiltshire.

Vallance, H.A. (1989) *The Great North of Scotland Railway*, New edition, Nairn.

Watson, N. (1987) *The Last Mill on the Esk: 150 Years of Papermaking*, Edinburgh.

Weatherill, L. (1974) *One Hundred Years of Papermaking. An illustrated history of the Guard Bridge Paper Co. Ltd., 1873–1973*, Guardbridge, Fife.

White, W. (1874) *History, Gazetteer and Directory of Suffolk*, Sheffield.

Williams, T.I. (1982) *A Short History of 20th Century Technology c. 1900–c. 1950*, Oxford.

Epilogue

Forty Years On

At the time C. Davidson and Sons was acquired in 1953 by British Plaster Board Ltd (now BPB plc) there were over 1,000 people employed. The mill was running on three shifts with one board and two paper machines making liner for plasterboard, packaging papers and a waterproof industrial paper, an associated finishing department, a machine-made paper bags department, a box factory and the multiwall paper sacks subsidiary of Abertay. In Aberdeen itself were the local branches of The Northern Waste Paper Company Ltd and of Davidsons Paper Sales Ltd.

The operations within the Davidson empire though, extended well beyond the boundaries of the Dee and the Don. To meet the mill's needs for its main raw material there were further waste paper branches in Dundee, Edinburgh, Glasgow and Newcastle-on-Tyne. Also in England were paperboard converting companies making cartons, solid fibreboard and rigid boxes, tubes, canisters and waxed paper; and covering the whole of the British Isles were a further six branches of Davidsons Paper Sales. Much further afield in Southern Rhodesia was Allied Rhodesian Manufacturers (Private) Ltd with its mill in Umtali making newsprint, cartonboard and plasterboard liner for the company's gypsum plants in southern Africa.

In contrast to the mode of operation in the 1990s when most support services are contracted out, there were then many more trades and professions on site at Mugiemoss. In addition to fitters, turners, electricians, all with their mates, there were plumbers, tilers, carpenters, catering staff, a nurse, transport drivers . . . and all were employed by Davidsons. Apprenticeships were also thriving. There was no Head Office as we know it today since Colonel Peter Davidson, then Chairman and Managing Director, strongly believed in a devolved and autonomous style of management.

It was recorded in the parent company's annual report and accounts for 1953/54, 'A full year of profits from C. Davidson & Sons Limited is, of course, included and I would like to mention here that the results achieved by that company have been most gratifying.' Having made such an auspicious start within the BPB family of companies, Davidson's has maintained this early reputation through timely investment, innovation, and dedication of its employees.

During the 1950s Mugiemoss played its part in the post World War II recovery taking place throughout the country. The development of many new towns like Harlow, Basildon, Skelmersdale and Livingston, around the major cities, gave rise to an increasing demand for liner for plasterboard. At that time the product was predominantly used for ceilings and the market was serviced by the Group's board plants, including the Gyproc operations in Shieldhall, Glasgow. Increased market penetration through property repairs and dry lining of walls was to come later. Following acquisition by BPB, therefore, the investment plans for the mill were quickly put into action. Production was expanded in most departments, with a focus on No. 4 machine. The drying section of the board machine was extended from 75 to 104 cylinders thereby increasing output from 4 to 5.5 tonnes per hour. As part of the £1.5 million investment, a new water treatment plant was commissioned and a viable transport fleet was established. Between 1959 and 1962 a new wet end comprising 8 vats, a new press section and new drying hood were installed.

At this time, good papermaking was as much an art as a science. The efficient running of the machines was highly dependent on the skills of the operators, supervisors and the papermaker in the absence of the sophisticated automatic control systems available today. To obtain and maintain the desired properties of the paper and board at the dry end of the machines regular testing of samples from the reels in the laboratory was required. A crucial part of the process was the regular adjustment of the furnish comprising both waste paper and virgin pulp. The wet end stock had to feel and look right, its good formation on the wires being achieved through manual controls, and optimum pressing and drying conditions were set by experience. Subsequent slitting, re-reeling and sheeting settings were made manually.

In the office the typewriters were entirely manual, paper sizes were quarto and foolscap, copies were made through the use of carbon paper and the mechanical calculators were operated by hand. Product weight was expressed in tons, hundredweights, quarters and pounds; and price was in pounds, shillings and pence. Multiplying weight by price to obtain sales values was therefore complex, but so easily accommodated had today's computers been available. Punched card systems were the vogue at that time and were used in the main for administrative duties like the calculation of wages – paid weekly in cash.

In the early 1960s it was recognised that there was a growing requirement for an uncoated cartonboard based on waste paper with a bleached white surface layer made from virgin pulp. Also at this time, sales of plasterboard were continuing to rise not only in the UK but also at BPB's associated companies in France and Belgium. Plans were therefore made to meet the growing needs of both these markets by installing a state-of-the-art Inverform paper machine, a British invention. The prototype was developed by St. Anne's Board Mill, Bristol and the second such machine for the UK was built for Thames Board at Purfleet; Davidson's was number three and this was commissioned in 1965 together with a new steam and power plant. This major investment, manufactured in Britain by Walmsleys of Bury and costing some £2.25 million, almost doubled the size of the mill overnight. No. 5 machine soon developed into the company's flagship and through a continuous development programme it became the leading board mill in the industry. Recent additions and modifications have assured its continuing status as one of the most productive paperboard machines in the world, and is now rated with an annual capacity of 130,000 tonnes, over three times its original capability. On start-up, imperial units of measurement were the norm, and still are the preferred units in the USA today. Both ivory and grey plasterboard liner were at that time 26 thousandths of an inch in thickness and the basis weights were both 160 pounds double crown per 480 sheets. The metric equivalents were soon adopted (650 microns in thickness and 390 grammes per square meter in weight) and few changes in specification were made until the late 1970s. Having successfully introduced Davidson's brand of white lined chipboard, 'Bucksboard', into the UK market, demand for plasterboard liner became so great that withdrawal from that part of the cartonboard sector had to be made, only to be introduced again in relatively small quantities for a few years from 1988.

It was also in 1965 that the Davidson Radcliffe Group Ltd (DRG) was formed to bring under one umbrella all the paper and packaging interests of BPB in the UK. All the operations though, continued to trade under their own well-known and respected names. There was, however, a little confusion in the industry for a time in that DRG and an associated logo was also widely used by The Dickinson Robinson Group, but clarity was restored when within BPB we became known as DRL. Two years later, in 1967, the stationery, box and carton making operations of C. Davidson and Sons (Packaging) Ltd together with other similar converters within the group formed Landor Cartons with its headquarters in Birmingham.

The 1970s were turbulent years for the industry. The rock steady leadership of both Colonel Peter Davidson and Harold Pearson, who had been appointed Managing Director in 1969, moulded the company to meet the ever changing circumstances. At the end of the decade Davidsons emerged in a far stronger position. Arising from the Arab/Israeli war in the Middle East, energy prices quadrupled, a three-day week was introduced as an energy saving measure, inflation really took hold, and to combat the effects of the cyclical nature of the paperboard industry, rigorous cost saving measures were introduced. This environment though, together with the rapid development of microchip based control technologies, provided the stimuli to computerise board manufacture and change the way the mill was run. Continuous four shift operation was introduced to accommodate the reducing normal hours worked per week, manning levels were trimmed, quality improved and outputs increased, all contributing to enhanced productivity and improved competitiveness.

In common with manufacturing industry in general, rapid expansion in all support services and departments at Mugiemoss was taking place. Fresh technologies were being developed, new markets and applications required servicing, cost and accounting procedures were becoming more rigorous, and many new management techniques needed to be introduced. The efficient running of a progressive company was therefore becoming more complex, needed more, well educated staff and naturally offices and laboratories in which to house them. New offices were completed in 1978 along two adjacent sides of Mugiemoss House, the original home of the Davidson family. These were a welcome replacement for the many green-painted, World War II wooden huts, which had been brought onto the site as a temporary measure some 30 years previously! During construction of the offices a forceful reminder came to light that this part of the 70 acres at Mugiemoss was a terminal moraine left over from the last ice age. This provided sound foundations for the most part but special footings were needed at the corner of the 'L' shaped building at the edge of the moraine. Settlement cracks at first floor level continue to be monitored.

The company was honoured to have HRH Prince Philip, Duke of Edinburgh visit Mugiemoss to open these offices formally and to see the many other developments then taking place in the works. Among these were the first stages of the all-embracing effluent water treatment plant in operation today. Although it was appreciated by only a few at the time, this event also marked the 25th anniversary of the company being part of BPB and was to be the first of further visits by members of the Royal Family.

Throughout Europe traditional markets were changing fast, especially in the packaging field, through product substitution and improvements in distribution methods. The traditional counter, bread and carrier bags made from paper were progressively being replaced by their polythene equivalents. In 1972, therefore, Davidsons opened a plastics division in the vacated Chambers fibreboard case factory on the Team Valley Trading Estate in Gateshead. A film extruder was installed and one of the first polythene recycling plants for post consumer waste was successfully commissioned. The promising life of the new venture, however, was cut short by the first 'oil crisis' because of multi-fold price increases on progressively more difficult to obtain raw materials.

In the paperboard industry older, less productive plant was being closed as part of sector-wide consolidation programmes and Mugiemoss was included in these changes. Production of 'Ibeco', the patented bitumen impregnated waterproof paper used in the construction and packaging industries, was transferred to No. 3 paper machine and No. 2 was closed and dismantled in 1978. Among ancillary plant that had also come to the end of its useful life in meeting the needs of a declining market was the paper creping machine. As demand for packaging papers and 'Ibeco' declined further in favour of plastic alternatives, the running of No. 3 machine progressively became less tenable and so this part of the mill was closed down in March 1981. This was a significant step for

Davidsons as it brought to an end, after 185 years, the making of paper at Mugiemoss. The sentimental ties were broken, allowing for clear focus on just boardmaking for the first time. The buildings were renovated and for a time housed a pallet-making and a paper-board core winding facility. The cores, or reel centres as they are known, were used by both Davidsons and other mills on the River Don. Further rationalisation was also under way in response to the rapidly changing markets in the flexible packaging field. Davidson's Paper Sales and the bag making operations at Mugiemoss were amalgamated with the company's related operations of O. J. Bradbury & Son, which had been acquired in 1959, to form Davidson Packaging based at Ruddington, near Nottingham. In contrast, sales from Nos 4 and 5 machines were going from strength to strength, not least as a result of much rationalisation taking place in the unlined chipboard sector in northern Europe.

Throughout the 1960s No. 4 machine was achieving an output of around 8 tonnes per hour, but operating efficiency gradually deteriorated in the 1970s when attention was directed towards the development of No. 5 machine. In 1978 plans for rebuilding No. 4 were updated and work began on erecting a new machine house alongside the existing one in the Spring of 1979. As new equipment was being installed the old No. 4 machine continued to run until 28 February 1980. Parts of the old machine, including 74 drying cylinders, were refurbished, and the new No. 4 machine had a superb start up making first quality board only 73 days later on 8 May. The BRDA formers and new press section performed beyond expectations. At the time, the changes were regarded as being 'heroic'. They set new standards for the board-making industry.

The progressive evolution of No. 5 machine is a text-book case of early and highly successful application of new technologies leading to ever increasing outputs, product development and acceptance in the market place world-wide for the mill's plasterboard liner. The first major change to the configuration of the five-station Inverform wet end was the addition in 1974 of the first Arcuforma in the world, made by the Finnish company, Tampella. This unit laid down a thicker layer of backs stock, permitting the machine to run faster and therefore increased its capacity. A measure of the success of this bold development was that the unit gave sterling service over the next 15 years. Davidsons were again breaking new ground in 1978 by installing (at a cost of some £1 million) computerised process control systems from two different suppliers, Measurex and Accuray, at the same time. Although it was appreciated at the time that this would lead to significant increases in productivity, it totally transformed the approach to papermaking over the next few years. Art had been largely replaced by science. Increased control over the process enabled a long term programme of lightweighting, enhanced quality and output increases to begin. In 1978 the first steps were taken, with very close co-operation of the gypsum plants in Britain and abroad, to reduce progressively the basis weight of plasterboard liner from 375 grammes per square metre (g/m^2). So great have the benefits been to the company that the area price of liner today is almost the same as twenty years ago, despite the value of sterling having depreciated by a factor of three over that time. Ivory plasterboard liner could now be made from a 100 per cent waste paper furnish. Additional benefits have also accrued to the plasterboard plants, enabling them to increase significantly the speed of their production lines. Further gains continue to be made, and 200 g/m^2 is now the standard - for the time being. The mill has always had an admirable reputation for being out there in front, and when it comes to awards it is there with the best. In relation to No. 5 machine, the Director General of the Confederation of British Industry, Sir John Methven, presented the mill in 1979 with its 'Ideas at Work' Award for Excellence in recognition of outstanding innovation and achievement in Scottish Industry.

Throughout the 1980s under the stewardship of Chris Bushell, who joined the company in 1976 as Deputy Managing Director, high utilisation rates were achieved on both board machines as their capacities rose incrementally. Five shift working was successfully introduced in 1985 and a five-year programme for rebuilding much of No. 5 machine was put into effect. The first stage in completely moving away from the Inverform concept comprised the timely installation of a fourdrinier unit to lay down the all-important gypsum bond (backs) ply. This was followed in 1989 with a middles ply fourdrinier and a mini-fourdrinier station for the top ply. Other plant enhancement included stock preparation, the size press, calender and a rebuild of the press section. These measures also enabled further reductions in basis weight of plasterboard liner to take place, while raising capacity and improving on quality even though waste paper fibre length and cleanliness were deteriorating.

Healthy growth in sales was maintained not only to Davidson Radcliffe's expanding converting activities in tubes, solid containers and board laminates, but also to third party customers including gypsum companies outwith Europe. Over 20 per cent of the company's output was being exported. In recognition of this, HRH The Duke of Kent, in his capacity as vice-chairman of the British Overseas Trade Board, visited the mill in 1988. The Queen's Award for Export Achievement was granted in 1989 and this accolade was repeated in 1992. Also in 1989 Davidsons gained yet another 'first': certification to BS 5750, the prestigious quality assurance registration. It was the first UK board mill to gain such an award, the first using recovered waste paper, the first high-volume producer, and the first mill in Scotland. This was followed by the company winning the National Training Award from the Enterprise Council, also in 1989.

In addition to the high reputation for its paperboard, cartons and multiwall paper sacks, Mugiemoss has also received a degree of notoriety for its publicity calendars. Landor Cartons' four-months-to-a-view with the perfect photographic trilogy of a sailing ship, a discretely attired lady and a landscape, was ably matched each year by Abertay's twelve pages of tasteful risqueness. Davidsons though, were setting another standard for specially commissioned water-colours of views, buildings and things very Scottish.

Improvements were continuously being made to the infrastructure of the Mugiemoss site to service the rising output and to meet the increasingly more stringent environmental requirements. By tightening up systems, much less process water from the river Don was being consumed. Also the consequent lower levels of effluent began to be more efficiently cleansed though the installation in 1986 of state-of the-art aerobic and anaerobic biological treatment plant at an initial cost of £0.75 million. Even the methane so generated continues to be used in a package boiler. Also in 1986 a brand new concept in waste paper treatment was put into practice when plant like a giant cement mixer was commissioned. Productivity in waste paper handling, treatment and cleaning was greatly enhanced, aided by this innovative 600 tonnes per day rotating soaking drum together with other sophisticated plant for removing the ever increasing quantities of plastics contraries. Changes were also taking place in the offices and in the area of site security.

The advances in sophisticated electronic equipment, and its rapidly falling cost, led to many commercial and administrative duties being carried out by personal computers, which soon became linked on local area networks and in the 1990s to full and comprehensive inter-site communications. The development of closed circuit television systems not only helped in the monitoring of remote and inaccessible parts of the manufacturing plant, but also led to the demise of a need for guard dogs and a team of security personnel. On the bank behind Mugiemoss House there are the graves of several of the company's well remembered and respected canine employees.

In 1985, when Harold Pearson was preparing to hand over the company's reins on his retirement to Chris Bushell, a major acquisition was under way. The Purfleet mill, Fiberite Packaging and the waste paper activities of the Thames subsidiary of Unilever came into the family in 1986. The extended Group's name was changed to Davidson Limited and the event was fully publicised on the company's impressive stand at Pakex, the international packaging exhibition, held at the NEC in Birmingham. 1986 also brought with it in September the tragic death at his Caskieben home of Peter Davidson, the last in the line of the original founding family.

Many at Mugiemoss were involved in assisting with the integration of the newly acquired operations down south and in laying down the foundations for their development over the next ten years. This turned out to be quite a natural process since Chris Bushell had been a member of the team at Purfleet when the Inverform machine was commissioned in the 1960s. Also many others from the Thames Group had joined Davidsons over the years. Mugiemoss was only continuing in its long established role of helping others to attain its own high standards using experience gained in Zimbabwe, the Netherlands and, closer to home, at Radcliffe.

With Chris Bushell taking on the additional responsibility for other non-gypsum companies within BPB, many of which were located in the North West of England, the Head Office of Davidsons Ltd moved from Mugiemoss in 1989 to new, purpose built offices at Gadbrook Park, Northwich. Since the centre of gravity was moving away from the previous Davidson family of companies, the name was changed in 1989 to BPB Paper & Packaging Ltd to reflect its membership of BPB and its main activities, and then to BPB Paperboard Ltd in 1996 when BPB itself took on a world-wide image and role. The mill at Mugiemoss therefore became known as BPB Davidson in 1989 and then BPB Paperboard, Davidson Mill, in 1996.

In bringing this short history up-to-date mention has to be made of the less than favourable trading conditions of the early 1990s. After the housing and construction boom of the 1980s a prolonged malaise was to follow, but Davidsons took advantage of the more difficult markets at home and expanded further its sales abroad, particularly to deep-sea markets both east and west. Demand for pulp, waste paper and paper products world-wide began to rise steeply in 1994 and prices fully reflected the boom which was to last for relatively few months into 1996. It was a timely reminder that the industry is highly cyclical and successful companies like Davidson need to remain fleet of foot to take full advantage of the changes in the market place both on the way up and coming down. The mill continued investing strongly in its machines, plant and support services, now under the chairmanship of John Goodall following Chris Bushell's retirement at the end of 1995. The finishing touches to the wet end of No. 5 machine were completed, the main drive capacity was increased and a new main winder installed. Further reductions in board basis weight and improvements to quality continued to be made. Plans have also been drawn up for further investment into the next century. These include a new warehouse and finished product logistics for the whole site and for the Group's gypsum wallboard plants.

The beneficial influence that Mugiemoss has had on the local community and further afield also needs recognition. Throughout its long history it has brought a measure of wealth into the Bucksburn area, many families have worked for several generations at the mill and the status of Aberdeen as a centre of excellence for papermaking has been enhanced. The presence of the Recreation Club has provided a centre for relaxation, indoor sports and more formal meetings for both Mugiemoss employees and for other groups in the area. The professionalism of its management committee has been tested on many occasions, not least of which was a fire which destroyed the main function hall in 1981. The club bar and lounge was duly back in operation the next day, but rebuilding the hall took a little longer. The playing fields have been progressively encroached upon for storing waste paper (which have also had their fires), but have long provided the venue for interdepartment and league cricket and football matches. Angling for salmon on the Mugiemoss reach of the river Don has been the favourite sport of many employees, not only for their own relaxation, but also to the benefit of the mill's many important customers. The mill continues to be an active member of the North East of Scotland section of the Paper Industry Technical Association (PITA) and it also maintains its long term relationship with the paper sciences department of Robert Gordon's University in Aberdeen. Davidsons has always appreciated that it has a duty in the paperboard industry to share its knowledge and experience with a wider audience. Its staff, therefore, continue to be regular contributors of articles to the trade and technical press, and presenters of papers at industry conferences. With intervals of a few years, many memorable Open Days have been held so that family members and dignitaries could really experience the characteristic aroma of wet waste paper, and see how that was being made into cardboard. The most memorable of these occasions was on May 28 1996 when HRH Princess Ann was the guest of honour at the events marking the mill's bi-centenary. Great importance has always been placed on Mugiemoss being a good neighbour in the community and that spirit, together with its assured future in the international paperboard market, will provide the foundation for the next 200 years as the mill welcomes in the 21st century.

There are but few industries which can match the history and evolution of the Davidson mill. Its continuing success can be summed up in the few words, 'Aye tae the fore'.

Index